Modern Algebra
An Introduction

Tim Anderson
University of British Columbia

Charles E. Merrill Publishing Company
A Bell & Howell Company
Columbus, Ohio

Merrill Mathematics Series
Erwin Kleinfeld, Editor

Published by
Charles E. Merrill Publishing Co.
A Bell & Howell Company
Columbus, Ohio 43216

International Standard Book Number: 0–675–08904–2

Library of Congress Catalog Number: 73–82414

AMS 1970 Subject Classifications: 12E05, 1201, 1401, 2001

1 2 3 4 5 6 — 79 78 77 76 75 74

Printed in the United States of America

Preface

This book is meant for a one term undergraduate course in algebra, assuming the student has mastered the calculus sequence.

The main topics of such a course are groups, rings, and fields. However, there is considerable variety in the way these topics are treated in texts, ranging from a strict axiomatic study of many algebraic structures to a fairly complete development of a few topics, notably Galois Theory.

In writing this book I have constantly tried to keep in mind that most people taking an introductory course in algebra will have very little intuition about the subject. For that reason I have included some topics from geometry in order to show why one studies groups and fields. In other words, the geometric sense we all have to some degree is called upon to nourish the algebraic sense.

As a consequence of this program, the book contains a chapter on elementary algebraic geometry, which is not a standard topic in a beginning algebra course. In many books used in a beginning algebra course, the important theory of unique factorization is introduced but is never developed and discussed to any great extent. However, in this book unique factorization is used in a familiar situation, the study of planar curves. This, incidentally, reflects the historical fact that the theory of curves, surfaces, and polynomial rings were quite intimately connected.

Another way in which this book differs from others is in its omission of Galois Theory, because this theory, which builds one algebraic structure upon another, requires a highly developed algebraic intuition and is properly dealt with in a graduate course. On the other hand, the beauty and importance of this theory compels one to say something about it, so we have chosen the historical route and have given Lagrange's theory of equations to show what was known about groups and equations in the days of Galois and Abel.

The following table shows the connections between the algebraic and geometric concepts discussed in the text.

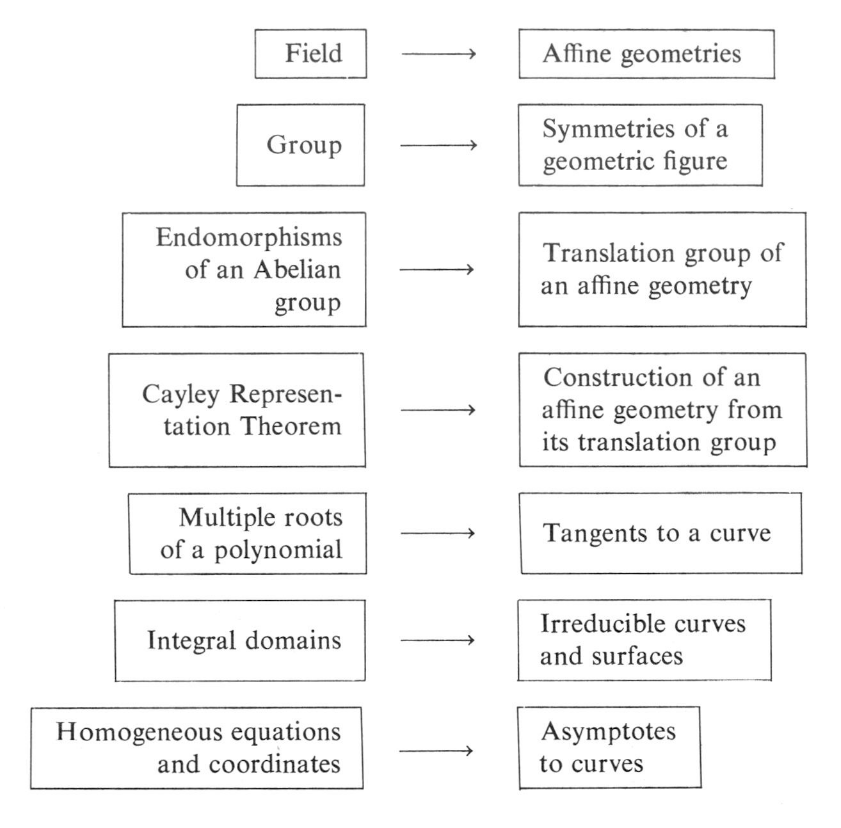

The approach used in this text should appeal not only to the student anticipating advanced study of mathematics in graduate school, but also to the student planning a career teaching mathematics on the secondary level.

I would like to thank Miss Mary Daisley for helping with the typing of the manuscript. Also, I am grateful to Professor Larry Roberts for his critical reading and suggestions about part of this book.

Contents

1

The Theory of Numbers

1-1 The Division Algorithm

Throughout this book we shall use some fairly common notation about sets, sums, and products. Most readers will already be familiar with these notations. However, for the convenience of those who are not familiar with them or who need a review, we have included in the Appendix a brief discussion concerning these matters.

The whole theory of divisibility that we shall discuss in this chapter is based on one of the most common facts of mathematics. This fact is simply that long division by integers is possible. Anyone with a minimum of schooling is aware of this. Yet, we shall constantly use this fact to gain real insight about integers, as well as abstract algebraic systems.

The division algorithm. If a and b are integers and $a > 0$, then there exist integers q and r such that

$$b = aq + r$$

where $0 \leq r < a$.

The integers q and r are called the *quotient* and *remainder* respectively.

Example 1-1. As an illustration of the utility of this algorithm, let us prove the following result: $n^5 - n$ is a multiple of 5 for any integer n.

Applying the division algorithm to the integers n and 5, we can find a quotient q and remainder r such that $n = 5q + r$, where $0 \leq r < 5$. Expanding $n^5 - n$ we have, after rearranging terms,

$$\begin{aligned} n^5 - n &= 3125q^5 + 3125q^4r + 1250q^3r^2 + 250q^2r^3 + 25qr^4 \\ &\quad - 5q + (r^5 - r) \\ &= 5(625q^5 + 625q^4r + 250q^3r^2 + 50q^2r^3 + 5qr^4 - q) \\ &\quad + (r^5 - r) \end{aligned}$$

Thus, to show $n^5 - n$ is a multiple of 5, it is sufficient to show $r^5 - r$ is a multiple of 5. It would appear then that we have merely exchanged the original problem for one of the very same sort. However, there is an essential difference: *The integer* $r = 0, 1, 2, 3,$ *or* 4. Now, we just check that for $r = 0, 1, 2, 3$, and 4, $r^5 - r = 0, 0, 30, 240$, and 1020 respectively. In each case, $r^5 - r$ is a multiple of 5.

An integer $a \neq 0$ is said to be a *divisor* of the integer b if there exists an integer n for which $b = an$. This is equivalent to saying the remainder r in the division algorithm is zero. In the case that a divides b, we write $a \mid b$; while if a does not divide b, we use the notation $a \nmid b$.

The following facts are obvious, but useful:

Proposition 1-1. Let a, b, and c be integers. Then: (a) If $a \mid b$ and $a \mid c$ then a divides $bx + cy$ for all integers x and y; and (b) if $a \mid b$ and $b \mid c$ then $a \mid c$.

Another very useful property of the integers is the so-called *Well Ordering Principle:*

If $\mathcal{C}$ is a collection of integers which contains at least one positive integer, then $\mathcal{C}$ contains a smallest positive integer. That is, there exists a positive integer $u \in \mathcal{C}$ such that $u \leq v$ for all positive integers v of $\mathcal{C}$.

It should be noted that the Well Ordering Principle does not extend to the real number system. The following set $\mathcal{C}$ clearly does not contain a smallest positive real number.

$$\mathcal{C} = 1, \frac{1}{2}, \ldots, \frac{1}{n}, \ldots$$

If a and b are nonzero integers, then any integer which divides both a and b is called a *common divisor* of a and b. A pair of integers a and b will have only a finite number of common divisors and the largest of these is called the *greatest common divisor* (G.C.D.) of a and b. If d is the greatest common divisor of a and b, we use the notation $d = (ab)$ to denote this fact. Since 1 divides every integer, we have $d \geq 1$. The integers a and b are called *relatively prime* if they have only trivial common divisors, i.e., if $(ab) = 1$.

Theorem 1-1. Let $d = (ab)$. Then, there exist integers x and y such that

$$d = ax + by$$

Proof: Consider the set $\mathcal{C}$ of integers of the form

$$ax' + by'$$

where x' and y' are integers. The set $\mathcal{C}$ contains the integers a and $-a$ since $a = a \cdot 1 + b \cdot 0$ and $-a = a(-1) + b \cdot 0$. Since either a or $-a$ is positive, at least one integer of $\mathcal{C}$ is positive. Therefore, because of the Well Ordering Principle, $\mathcal{C}$ contains a smallest positive integer d^* and

$$d^* = ax + by$$

for some integers x and y.

We shall show $d^* = d$. Since d divides both a and b, d divides d^*(Proposition 1-1); hence,

$$d^* \geq d$$

To prove $d = d^*$, we must show, therefore, that

$$d^* \leq d$$

and this will be accomplished by proving d^* is a common divisor of a and b.

Using the division algorithm, we have $a = qd^* + r$ for a quotient q and remainder r, where $0 \leq r < d^*$. Now, $r = a - qd^* = a - q(ax + by) = a(1 - qx) + b(-qy)$. Therefore, r is of the form

$r = ax' + by'$, where $x' = 1 - qx$ and $y' = -qy$ are integers. As r is smaller than d^*, it follows from the fact that d^* is the smallest integer in $\mathcal{C}$ that $r = 0$. That is, d^* divides a. A similar argument shows that d^* divides b. Hence, d^* is a common divisor of a and b, as required.

Remark 1-1. We have actually proved that the greatest common divisor of integers a and b is the smallest positive integer of the form $ax + by$, where x and y are integers.

Corollary 1-1. The greatest common divisor of the integers a and b is divisible by all the common divisors of a and b.

Proof: For the greatest common divisor $d = (a,b)$, we have $d = ax + by$ for integers x and y. Any other common divisor d' of a and b must divide $ax + by$; hence, it divides d.

1-2 The Euclidean Algorithm

The characterization (Theorem 1-1) of the greatest common divisor will be seen to be quite important in the theory of divisibility of integers. Before giving the ramifications of this, let us indicate how the greatest common divisor is found in practice.

Since the divisors of an integer are the same as the divisors of its negative, we may assume a and b are both positive integers when finding their greatest common divisor.

Dividing a into b,

$$b = aq + r \tag{1-1}$$

for a remainder r, $0 \leq r < a$, and quotient q.

If $r = 0$, then $a \mid b$ and $(a,b) = a$. If $r > 0$, then dividing r into a we have

$$a = q_1 r + r_1 \tag{1-2}$$

for a quotient q_1 and a remainder r_1, where $0 \leq r_1 < a$. What happens if $r_1 = 0$? In this case, we have $r \mid a$ by Equation (1-2); hence, by Equation (1-1), $r \mid b$ as well, i.e., r is a common divisor of a and b. Moreover, from Equation (1-1) we see that if $d = (a,b)$, then $d \mid r$, i.e., $r = d$ if $r_1 = 0$.

If $r_1 > 0$, we divide r_1 into r to get

$$r = q_2 r_1 + r_2$$

for a quotient q_2 and remainder r_2, $0 \leq r_2 < r_1$. In fact, we repeat this process until we reach a remainder which is zero. Thus, we arrive at a sequence of equations

$$\begin{aligned} b &= aq + r \\ a &= q_1 r + r_1 \\ r &= q_2 r_1 + r_2 \\ &\cdot \\ &\cdot \\ &\cdot \\ r_k &= q_{k+2} r_{k+1} + r_{k+2} \\ &\cdot \\ &\cdot \\ &\cdot \end{aligned} \tag{1-3}$$

which must ultimately produce a remainder which $= 0$, since the sequence $a > r > r_1 > \ldots > r_{k+2} > \ldots \geq 0$ cannot continue indefinitely. *The last nonzero remainder in this process is the greatest common divisor of a and b.* The proof that it does produce the greatest common divisor is accomplished by working backwards through the equations (1-3) and showing this last nonzero remainder is a common divisor. Then by going through the equations in reverse order, it is shown that the greatest common divisor of a and b divides this last nonzero remainder; whence, they are equal.

The algorithm described in the equations (1-3) is called the *Euclidean algorithm.*

Example 1-2. Carrying out the Euclidean algorithm for $d = (482,74)$ we find

$$\begin{aligned} 482 &= 6 \cdot 74 + 38 \\ 74 &= 1 \cdot 38 + 36 \\ 38 &= 1 \cdot 36 + 2 \\ 36 &= 18 \cdot 2 + 0 \end{aligned}$$

Thus, $2 = (482,74)$.

In Theorem 1-1, it was proved for the greatest common divisor $d = (a,b)$ that $d = ax + by$ for integers x and y. These integers are found by starting with the equation of the algorithm [equations (1-3)] in which the last nonzero remainder occurs and, then, working back from this equation, finally expressing the last nonzero remainder, after a series of substitutions, in terms of a and b.

In Example 1-2 we found that $2 = (482,74)$ after applying the algorithm

$$\underline{482} = 6\cdot\underline{74} + \underline{38} \tag{1-4}$$

$$\underline{74} = 1\cdot\underline{38} + \underline{36} \tag{1-5}$$

$$\underline{38} = 1\cdot\underline{36} + \underline{2} \tag{1-6}$$

$$36 = 18\cdot 2 + 0 \tag{1-7}$$

Here we have underlined the quantities essential for the substitution process in writing

$$2 = 482x + 74y$$

for integers x and y. To find the integers x and y, we know from Equation (1-6) that $2 = \underline{38} + \underline{36}(-1)$. With the aid of Equation (1-5), we may replace the $\underline{36}$ with $\underline{74}$ and $\underline{38}$, i.e.,

$$\begin{aligned} 2 &= \underline{38} + (\underline{74} - \underline{38})(-1) \\ &= \underline{38}(2) + \underline{74}(-1) \end{aligned}$$

Then, using Equation (1-4) we eliminate the $\underline{38}$ from the equation $2 = \underline{38}(2) + \underline{74}(-1)$. Thus, $2 = (\underline{482} - 6\cdot\underline{74})(2) + \underline{74}(-1) = 482(2) + 74(-13)$. Hence, we may choose $x = 2$ and $y = -13$.

1-3 Mathematical Induction

The Well-Ordering Principle states that if a set $\mathcal{C}$ of integers contains at least one positive integer, then it contains a smallest positive integer. With the aid of this principle, we formulate a method of proving theorems which is quite useful. This method is called the *Principle of Mathematical Induction*, which states the following:

Proposition 1-2. Let $P(n)$ be a statement about the positive integer n. Suppose: (a) $P(1)$ is a true statement; and (b) $P(k + 1)$ is true whenever $P(k)$ is true. Then the statement $P(n)$ is true for every positive integer n.

Before explaining why this proposition is valid, let us see what it says. As an illustration, suppose we have a sequence of cars and these cars are traveling too closely to one another. If car k in this sequence were to suddenly stop, so would car $(k + 1)$ (it would stop by crashing into car k). Clearly, if the very first car were to stop quickly, then every car in the sequence would stop.

The Mathematical Induction Principle simply states that such a chain reaction occurs in establishing the truth of the proposition $P(n)$. That is, if somehow the truth of the proposition $P(k)$ guarantees the truth of $P(k + 1)$, then $P(n)$ will be true for every positive integer n if it is true for the first integer 1.

Let us now give a proof of Proposition 1-2. Assume $P(n)$ is a statement satisfying (a) and (b). We want to show $P(n)$ is true for all positive integers n. Consider the set $\mathcal{C}$ of those positive integers for which $P(n)$ is *false*.

If there are any integers at all in $\mathcal{C}$, then there must be a smallest positive integer m in $\mathcal{C}$. Thus, $P(m)$ is false but $P(m')$ is true for all integers m' such that $m' < m$. Now, $m > 1$ because of condition (a). As $k = m - 1$ is, therefore, a positive integer which is smaller than m, $P(k)$ is true. But condition (b) shows that $P(k + 1)$ is true, i.e., $P(m)$ is true. However, this contradicts the choice of the integer m. Therefore, the set contains no integers at all. In other words, $P(n)$ is true for every positive integer n.

Often empirical evidence leads one to make a conjecture $P(n)$ about integers n, and induction is used to actually prove the conjecture. For example, one might notice that the product of three consecutive integers always seems to be divisible by 6. To prove this, one considers the following statement:

> $P(n)$: "For each positive integer n, $n(n + 1)(n + 2)$ is a multiple of 6."

For $n = 1$, $n(n + 1)(n + 2) = 1 \cdot 2 \cdot 3 = 6$; hence, $P(1)$ is true. This verifies condition (a) for the statement $P(n)$.

We have to show now that condition (b) holds for our statement. Assume then that $P(k)$ is true, i.e., $k(k + 1)(k + 2)$ is divisible by 6. Is $P(k + 1)$ true? To answer this, it is necessary to look at $(k + 1)(k + 2)(k + 3)$. Now in any case, we have the identity $(k + 1)(k + 2)(k + 3) = k(k + 1)(k + 2) + 3(k + 1)(k + 2)$. As either $k + 1$ or $k + 2$ is even, $3(k + 1)(k + 2)$ is a multiple of 6. Therefore, $(k + 1)(k + 2)(k + 3)$ is a multiple of 6, i.e., $P(k + 1)$ is true. Proposition 1-2 now shows that $P(n)$ is true for all positive integers n.

As another example of proving theorems by induction, let us show for a positive integer n that

$$1 + 2 + \ldots + n = \frac{n(n + 1)}{2} \tag{1-8}$$

If $n = 1$, then the right-hand side of Equation (1-8) is $1(1 + 1)/2$, which is the left-hand side of Equation (1-8); hence, Equation (1-8) is valid for $n = 1$.

Assuming Equation (1-8) is true for $n = k$, then for $k + 1$ we have

$$\begin{aligned} 1 + 2 + \ldots + (k + 1) &= (1 + 2 + \ldots + k) + (k + 1) \\ &= \frac{k(k + 1)}{2} + (k + 1) \\ &= \frac{(k + 1)(k + 2)}{2} \end{aligned}$$

Thus, Equation (1-8) is also true for $n = k + 1$. Therefore, by induction the formula [Equation (1-8)] is valid for all positive integers n.

Let us remark that the Induction Principle may be formulated in a slightly different way.

Proposition 1-3. Let $P(n)$ be a statement about the integer n. Suppose (a) $P(n_0)$ is true for some positive integer n_0 and (b) $P(k)$ is true providing $P(m)$ is true for all m, $n_0 \leq m < k$. Then $P(n)$ is true for all $n \geq n_0$.

1-4 Primes and Unique Factorization

An integer $p > 1$ which cannot be written as a product of two smaller positive integers is called a *prime*. Thus, 2,3,5,7,11, . . . are primes. As we shall see throughout this book, primes play a basic role in many algebraic studies. In regards to integers themselves we have the following theorem:

Theorem 1-2. *The Fundamental Theorem of Arithmetic.* Any integer > 1 is either itself a prime or it can be factored uniquely as a product of primes.

Before given the proof of Theorem 1-2, we require a preliminary result.

Theorem 1-3. If an integer a divides the product bc of integers b and c and if $(a,b) = 1$, then a divides c.

Proof: If $(a,b) = 1$, there are integers x and y such that $1 = ax + by$. (See Theorem 1-1.) Multiplying this equation by c, we have $c = acx + bcy$. But $acx + bcy$ is divisible by a; hence, c is also divisible by a.

Corollary 1-2. If a prime p divides the product bc of integers b and c, then either p divides b or p divides c.

Proof: The greatest common divisor (p,b) is either p or 1 because p is prime. If $(p,b) = p$, then p divides b. If $(p,b) = 1$, then p divides c because of Theorem 1-3.

Now, we may prove the Fundamental Theorem of Arithmetic. This theorem makes two assertions. It states: (1) for each integer $n > 1$, there are primes $p_1, \ldots, p_s$, $1 \leq s$ (not necessarily distinct primes) such that $n = p_1 \ldots p_s$; (2) if also $n = q_1 \ldots q_r$, where $q_1, \ldots, q_r$ are primes, then the p_i's and q_i's are identical.

We prove (1) using induction, i.e., Proposition 1-3. If every integer less than k is either a prime or a product of primes, and if k is not prime, then $k = ab$ for positive integers a and b, where $a < k$ and $b < k$. Each a and b is a product of primes, which implies that k is a product of primes. Taking $n_0 = 2$ in Proposition 1-3, it follows from that proposition that every integer > 1 is either prime or a product of primes.

To verify (2), suppose $n = p_1 \ldots p_s$ and $n = q_1 \ldots q_r$ for primes $p_1, \ldots, p_s, q_1, \ldots, q_r$. Then, $p_1 \ldots p_s = q_1 \ldots q_r$, which shows that p_1 divides the product $q_1 \ldots q_r$. From Corollary 1-2, it follows that p_1 divides some q_j. However, q_j is prime; hence, $p_1 = q_j$. Then, canceling the prime p_1 from both sides of the equation $p_1 \ldots p_s = q_1 \ldots q_r$, and repeating the preceding argument for each of the remaining primes $p_2, \ldots, p_s$, we will find that the primes $p_1, \ldots, p_s$ are identical, in some order, to the primes $q_1, \ldots, q_r$.

A *rational* number is one which can be represented as the ratio of two integers. A real number which is not rational is called *irrational*. As an application of the uniqueness part of the Fundamental Theorem of Arithmetic, we prove the following:

Theorem 1-4. If p is a prime, then $\sqrt{p}$ is irrational.

Proof: Suppose to the contrary that $\sqrt{p}$ is rational. Then, there are integers m and n such that $\sqrt{p} = m/n$ or

$$n^2p = m^2 \tag{1-9}$$

Now, if the prime factorization of n is $n = p_1 \ldots p_r$, where $p_1, \ldots, p_r$ are primes, and if $m = q_1 \ldots q_s$, where $q_1, \ldots, q_s$ are primes, then $n^2 = p_1 \ldots p_r p_1 \ldots p_r$ and $m^2 = q_1 \ldots q_s q_1 \ldots q_s$. In particular, n^2 and m^2 are products of an even number of prime factors (namely the product of $2r$ and $2s$ prime factors respectively). Looking at Equation (1-9) we see that the right-hand side of this equation is the product of an even number of prime factors, while the left-hand side is the product of an odd number $(2r + 1)$ of prime factors.

Hence, Equation (1-9) violates the unique factorization part of the Fundamental Theorem of Arithmetic. Therefore, $\sqrt{p}$ cannot be rational.

If the distinct prime factors of an integer a are $p_1, \ldots, p_n$, then

$$a = p_1^{e_1} \ldots p_n^{e_n} \tag{1-10}$$

where $e_1, \ldots, e_n$ are integers ≥ 1. From the factorization [Equation (1-10)], it is straightforward to list the divisors of a. In particular, if the prime factorization of two integers a and b is known, then their greatest common divisor can be found by inspection.

Generally, if $a_1, \ldots, a_k$ are nonzero integers, we let $d = (a_1, \ldots, a_k)$ denote the greatest common divisor of these integers. The number d has the property that it is divisible by any integer which divides each of $a_1, \ldots, a_k$.

In contrast to common divisors, a *common multiple* of integers $a_1, \ldots, a_k$ is an integer m for which $a_1 \mid m, \ldots, a_k \mid m$. The least common multiple of $a_1, \ldots, a_k$ is denoted by $[a_1, \ldots, a_k]$.

1-5 Distribution of Primes

There are many important questions concerning primes which are as yet unsolved. For example, there is *Goldbach's Conjecture*, which asks whether every even integer is the sum of two primes. Other interesting questions involve the distribution of primes. For example, there are exactly 8 primes in the interval from 1 to 20, but only 4 primes between 20 and 40. Can one predict how many primes there are in a given interval of integers?

Proposition 1-4. If n is an integer ≥ 1, then there is a prime p such that $n < p \leq n! + 1$.

Proof: Consider the integer $N = n! + 1$. If N is prime, we may take $p = N$. If N is not prime, it has some prime factor p. If $p \leq n$, then p would divide $n!$; hence, p would divide $N - n!$, which is ridiculous since $N - n! = 1$. Therefore, $n > p$.

As a corollary to this we have the following theorem:

Theorem 1-5. *Euclid's Theorem.* There are infinitely many primes.

There are long sequences of consecutive integers which are barren of primes, as the next result shows.

Proposition 1-5. For an integer $n \geq 2$, there are no primes between $n! + 2$ and $n! + n$.

Proof: Indeed, as $n!$ is the product of all the integers between 1 and n, 2 divides $n! + 2$, 3 divides $n! + 3, \ldots, n$ divides $n! + n$.

The celebrated *Prime Number Theorem* states that if $\Pi(x)$ denotes the number of primes $\leq x$, then

$$\Pi(x) \backsim \frac{x}{\log x} \tag{1-11}$$

in the sense that the ratio of $\Pi(x)$ and $x/\log x$ tends to 1 as x tends to infinity.

It seems incredible that the distribution of primes can be described so succinctly as in (1-11). It is well beyond the scope of this book to explain the Prime Number Theorem. Such a result is dealt with in a course on *analytic number theory.* However, we shall give some indication how the logarithm function enters in the distribution of primes.

By induction it can be readily shown that

$$1 + x + \ldots + x^n = \frac{1 - x^{n+1}}{1 - x}$$

Now if $|x| < 1$, then as n tends to infinity, $(1 - x^{n+1})/(1 - x)$ tends to $1/(1 - x)$, i.e.,

$$\lim_{n\to\infty} \frac{1 - x^{n+1}}{1 - x} = \frac{1}{1 - x} \quad \text{if } |x| < 1$$

In calculus, one usually writes

$$\sum_{n=0}^{\infty} x^n = \lim_{n\to\infty} (1 + x + \ldots + x^n)$$

Thus, for the geometric series $\sum_{n=0}^{\infty} x^n$ we have

$$\sum_{n=0}^{\infty} x^n = \frac{1}{1 - x} \quad \text{if } |x| < 1 \tag{1-12}$$

Now, consider the product

$$\prod_{p \leq x} \frac{1}{1 - p^{-1}} \tag{1-13}$$

where this product includes all primes p which are $\leq x$.

Euler observed that

$$\prod_{p \leq x} \frac{1}{1 - p^{-1}} > \sum_{m \leq x} \frac{1}{m} \tag{1-14}$$

where the summation includes all integers m such that $m \leq x$. To understand this, note for a prime p that $|p^{-1}| < 1$, so by Equation (1-12)

$$\frac{1}{1 - p^{-1}} = 1 + \frac{1}{p} + \left(\frac{1}{p}\right)^2 + \ldots + \left(\frac{1}{p}\right)^n + \ldots$$

Thus,

$$\prod_{p \leq x} \frac{1}{1 - p^{-1}} = \left(1 + \frac{1}{2} + \ldots + \left(\frac{1}{2}\right)^n + \ldots\right) \cdots \left(1 + \frac{1}{q} + \ldots + \left(\frac{1}{q}\right)^n + \ldots\right) \tag{1-15}$$

where q is the largest prime which is $\leq x$. If m is any integer $\leq x$, then m is the product of primes $\leq x$. Therefore, $1/m$ is the product $1/m = (1/p_1) \ldots (1/p_r)$ where $p_1, \ldots, p_r$ are primes $\leq x$. By multiplying out Equation (1-15), we will actually get the term $1/m$ (plus many others). For example, if $x = 50$, then $1/21 = (1/3)(1/7)$, and in the product $\prod_{p \leq 50} 1/(1 - p^{-1})$ we will have the factors $1/(1 - 3^{-1})$ and $1/(1 - 7^{-1})$, whose product

$$[1/(1 - 3^{-1})][1/(1 - 7^{-1})] = 1 + 1/7 + 1/3 + (1/3)(1/7) + \ldots$$

Let $[x]$ be the largest integer which is $\leq x$. Then, $\sum_{m \leq x} 1/m = \sum_{m=1}^{[x]} 1/m$, and comparing this sum with the area under the graph (see Figure 1-1) of $y = 1/t$ for $0 \leq t \leq [x] + 1$, we have

$$\sum_{m \leq x} \frac{1}{m} > \int_1^{[x]+1} \frac{dt}{t} > \int_1^x \frac{dt}{t}$$

Hence, as the last integral is $\log x$, we have from Equation (1-14) that

$$\prod_{p \leq x} \frac{1}{1 - p^{-1}} > \log x \tag{1-16}$$

It is by such estimates that the logarithm function enters into questions about the distribution of primes.

[The sum of the areas of the shaded rectangles in Figure 1-1 is $1 + (1/2) + \ldots + [x]$, which equals $\sum_{m=x} (1/m)$.]

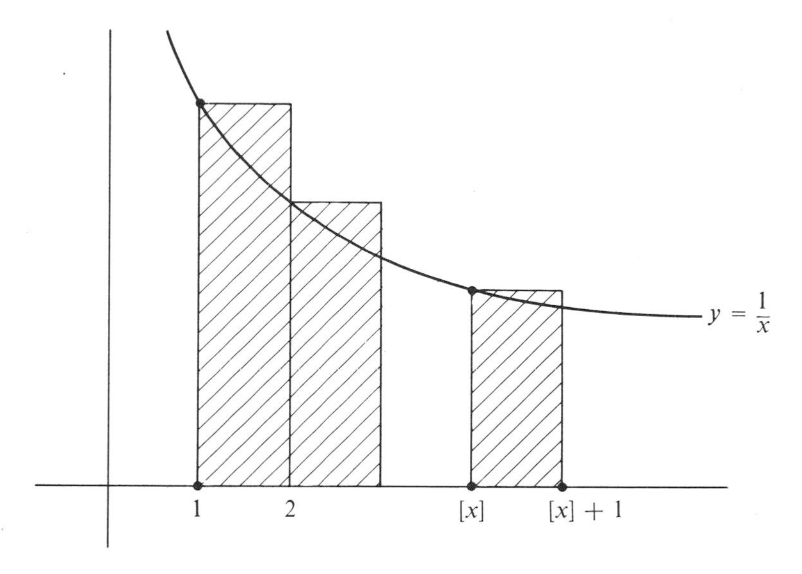

Figure 1-1

1-6 Diophantine Equations

For given integers a, b, and c, we ask: When does the equation

$$ax + by = c \tag{1-17}$$

have a solution in *integers* x and y? Such an equation is called a *Diophantine* equation. Of course, there will be many *rational* solutions for the numbers x and y; but in the study of Diophantine equations, one is interested only in the integral solutions of the equation.

There are many ways in which Diophantine equations arise. For example, if a theater normally sells tickets at \$1.85 each but admits students for \$1.25, and one night the theater collects \$485 in admissions, one can ask: How many regular admissions and student admissions were sold? To answer this question, one must solve the equation $1.85x + 1.25y = 485$ for integers x and y, or equivalently, solve $185x + 125y = 48{,}500$.

For the Diophantine equation [Equation (1-17)], we have a complete answer regarding its solvability.

Theorem 1-6. For integers a, b, and c, the equation

$$ax + by = c \tag{1-18}$$

has a solution in integers x and y if and only if (a,b) divides c.

Proof: For integral solutions x and y of Equation (1-18) the left-hand side of that equation will be a multiple of any common divisor of a and b; hence, (a,b) is a divisor of c.

Conversely, suppose the greatest common divisor $(a,b) = d$ is a divisor of c. According to Theorem 1-1, there are integers x_0 and y_0 for which $d = ax_0 + by_0$. Multiplying this equation by the integer c/d, we find $c = a(cx_0/d) + b(cy_0/d)$. Therefore, the integers $x_1 = cx_0/d$, $y_1 = cy_0/d$ provide a solution of Equation (1-18).

Equation (1-18) will have a number of integral solutions if it has any at all. To compare the solutions suppose we have another solution x and y in addition to the solution x_1,y_1 we have already found. Then, we will have

$$ax_1 + by_1 = c$$

and

$$ax + by = c$$

Subtracting these equations, $a(x - x_1) + b(y - y_1) = 0$, or $a(x - x_1) = b(y_1 - y)$. Now canceling out the greatest common divisor $d = (a,b)$,

$$\frac{a}{d}(x - x_1) = \frac{b}{d}(y_1 - y) \qquad \textbf{(1-19)}$$

It follows from Equation (1-19) that a/d divides the product $(b/d)(y_1 - y)$. However, $(a/d,b/d) = 1$. (See Exercise 1-4, Problem 4.) Therefore, a/d divides $y_1 - y$, or

$$y_1 - y = t\frac{a}{d} \qquad \textbf{(1-20)}$$

for some integer t. Then, substituting Equation (1-20) into Equation (1-19), we find

$$x - x_1 = t\frac{b}{d} \qquad \textbf{(1-21)}$$

Therefore, the general solution x,y of Equation (1-18) is given in terms of the solution x_1,y_1 by the equations

$$\begin{aligned} x &= x_1 + t\frac{b}{a} \\ y &= y_1 - t\frac{a}{d} \end{aligned} \qquad \textbf{(1-22)}$$

where t is an integer. In fact, by substituting Equation (1-22) into Equation (1-18), it may be verified that Equation (1-22) defines a solution for *all* integers t.

Let us note the following:

Proposition 1-6. For a prime p the Diophantine equation

$$px + by = c$$

always has a solution providing that $p \nmid b$.

Exercises

Exercise 1-1

1. Show that the integers q and r of the division algorithm are unique.
2. Show that an integer is divisible by 9 if and only if the sum of its digits is divisible by 9.
3. Derive the division algorithm from the Well Ordering Principle.
4. Show that if $d = (a,b)$, then the integers a/d and b/d are relatively prime.
5. Show that if a, b, and c are integers such that $a \mid c$, $b \mid c$, and $(a,b) = 1$, then $ab \mid c$.
6. Prove that integers x and y of Theorem 1-1 are relatively prime.
7. Show that if a and b are relatively prime positive integers, then all but a finite number of positive integers n can be written in the form $n = ax + by$, where x and y are nonnegative integers.
8. Show that if a, b, and c are integers such that $(a,c) = (b,c) = 1$, then $(ab,c) = 1$.

Exercise 1-2

1. Calculate the greatest common divisor of each of the following pairs of integers and express it in terms of the given integers as in Theorem 1-1.
   (a) (41,13)
   (b) (400,512)
   (c) (120,383)
2. Prove Theorem 1-1 from the Euclidean algorithm.
3. Suppose a and b are positive integers and $b = aq + r$ for a quotient q and a remainder r, where $0 < r < a$. Show that $(a,b) = (a,r)$. Then using this result explain why the greatest common divisor and a and b are the last nonzero remainder in the Euclidean algorithm for a and b.

Exercise 1-3

1. Give an example of a statement $P(n)$ which satisfies condition (a) but not (b) of Proposition 1-2.

2. Prove by induction that for a number x

$$1 + x + \ldots + x^n = \frac{1 - x^{n+1}}{1 - x}$$

3. Show by induction on the positive integer n that

$$(a + b)^n = \sum_{r=0}^{r=n} \binom{n}{r} a^r b^{n-r}$$

where

$$\binom{n}{r} = \frac{n!}{r!(n - r)!}$$

4. Using induction prove that

$$1^2 + 2^2 + \ldots + n^2 = \frac{n(n + 1)(2n + 1)}{6}$$

Exercise 1-4

1. Prove that the only positive integer n for which n, $n + 2$, and $n + 4$ are all primes is $n = 3$.
2. For positive integers m and n prove that $\sqrt[m]{n}$ is either an integer or irrational.
3. Show that rational numbers are repeating decimals and conversely (eg., $1/7 = .142857142857142857\ldots$).
4. Show that if p is prime and r is an integer, $1 \leq r \leq p - 1$, then the binomial coefficient $\binom{p}{r}$ is divisible by p.
5. Show that if $2^n - 1$ is prime for a positive integer n, then n is itself prime.
6. Show that if $2^n + 1$ is prime for a positive integer n, then n is a power of 2.
7. An integer m is said to be *perfect* if m is the sum of all its positive divisors $< m$. For example, $6 = 1 + 2 + 3$ and $28 = 1 + 2 + 4 + 7 + 14$ are perfect numbers. Show that if $2^k - 1$ is prime, then $2^{k-1}(2^k - 1)$ is a perfect number.
8. Prove by induction on the positive integer a that if p is a prime, then $a^p - a$ is divisible by p.
9. Show that for the least common multiple $[a,b]$ and greatest common divisor (a,b) of the integers a and b, $ab = (a,b)[a,b]$.
10. Let $f(x) = a_0 + a_1x \ldots + a_nx^n$ be a polynomial with integer coefficients. Show that for some positive integer k, $f(k)$ is not prime.
11. For a positive integer n, show that if n is not prime, then the smallest prime dividing n is no greater than $\sqrt{n}$. Using this, decide whether 3,997 is prime.

Exercise 1-5

1. Show that the highest power of a prime p dividing $n!$ is

$$\left[\frac{n}{p}\right] + \left[\frac{n}{p^2}\right] + \cdots$$

where $[x]$ is the largest integer $\leq x$.

2. Show that every prime p is of the form $4n + 1$ or $4n + 3$, where n is an integer.

3. Prove that if $p_1 < \ldots < p_k$ are primes of the form $4n + 3$, then there is a prime p of the form $4n + 3$ such that $p_k < p \leq 4p_1 \ldots p_k + 3$. (Hint: Show that the product of integers of the form $4n + 1$ is again of the form $4n + 1$. Then look at the prime factors of the integer $4p_1 \ldots p_k + 3$.)

4. Prove there are infinitely many primes of the form $6n + 5$.

5. Using the fact that $\log(1 + x) < x$ when x is positive, show that

$$\log \prod_{p \leq x} \frac{1}{1 - p^{-1}} < \sum_{p \leq x} \frac{1}{p} + \sum_{p \leq x} \frac{1}{p(p - 1)}$$

where the sums and product include all primes $p \leq x$.

6. Using the result of the preceding problem, show that

$$\log \prod_{p \leq x} \frac{1}{1 - p^{-1}} \leq \sum_{p \leq x} \frac{1}{p} + 1$$

by comparing

$$\sum_{p \leq x} \frac{1}{p(p - 1)}$$

with

$$\sum \frac{1}{n(n - 1)}$$

where the latter sum includes all integers n, $2 \leq n \leq x$. [Hint: Observe that $1/[n(n - 1)] = [1/(n - 1)] - (1/n)$.]

7. Show that

$$\sum_{p \leq x} \frac{1}{p} > \log(\log x) - 1$$

and conclude that

$$\lim_{x \to \infty} \sum_{p \leq x} \frac{1}{p} = \infty$$

Exercise 1-6

1. Show that the Diophantine equation

$$27x + 48y = 129$$

has an integral solution for x and y, then find all solutions of this equation. Find the smallest positive integral values of x and y which solve this equation.

2. If a, b, c, and n are integers, then when does the Diophantine equation

$$ax + by + cz = n$$

have an integral solution for x, y, and z?

3. Find all integral solutions of

$$6x + 9y + 33z = 30$$

2

Congruences and Abstract Number Systems

2-1 Congruences

For integers a, b, and n we write $a \equiv b \pmod{n}$ if $a - b$ is divisible by n. The relation $a \equiv b \pmod{n}$ is called a *congruence* and the integer n is called the *modulus* of the congruence. The modulus is always taken to be nonzero.

The integers a and b are congruent modulo n if, and only if, a and b leave the same remainder when divided by n. Indeed, applying the division algorithm to a and b, $a = qn + r$ and $b = q'n + r'$ for integers q, q', n, and n', where $0 \leq r < n$ and $0 \leq r' < n$. Hence, $a - b = (q - q')n + (r - r')$ and $a \equiv b \pmod{n}$ if and only if n divides $r - r'$. However, because r and r' are small relative to n, n divides $r - r'$ if, and only if, $r = r'$.

A congruence is much like an equation, as the following shows.

Proposition 2-1. If $a \equiv b \pmod{n}$ and c and d are congruent integers modulo n, then $a + c \equiv b + d \pmod{n}$ and $ac \equiv bd \pmod{n}$.

In other words, a congruence is unchanged when congruent quantities are added or multiplied to both sides of the congruence.

In verifying this proposition, we have $a - b = sn$ and $c - d = tn$ for integers s and t. Thus, $(a + c) - (b + d) = (s + t)n$, or $a + c \equiv b + d \pmod{n}$. Also, $ac - bd = (a - b)c + (c - d)b = (sc + tb)n$, and $ac \equiv bd \pmod{n}$.

On the basis of this simple proposition one can perform rather spectacular computations. For example, to find the remainder when 48^{72} is divided by 7, we note that $48 \equiv -1 \pmod{7}$; whence, $48^{72} \equiv (-1)^{72} \equiv 1 \pmod{n}$. So, the remainder must be 1.

Since a congruence is rather like an equation, it is natural to ask which congruences can be solved for an unknown. Generally, this is a deep and difficult problem. However, an answer can be given for those congruences which are of degree one in the unknown.

Theorem 2-1. The congruence $ax \equiv b \pmod{n}$ has a solution x if and only if (a,n) divides b.

Proof: For $ax \equiv b \pmod{n}$ if and only if there is an integer y such that $ax - b = yn$. But this is nothing more than the Diophantine equation $ax + y(-n) = b$, and we have seen before that a necessary and sufficient condition for its solvability is for (a,n) to divide b.

In the case of a prime p, the condition for solvability of $ax \equiv b \pmod{p}$ is much like that of an equation $ax = b$ in numbers because $(a,p) = 1$ or p. Thus, we have the result: The congruence $ax \equiv b \pmod{p}$ can be solved for x if $a \not\equiv 0 \pmod{p}$. Moreover, when $a \not\equiv 0 \pmod{p}$ the solution x is unique modulo p. For if also $ay \equiv b \pmod{p}$, then subtraction gives $a(x - y) \equiv 0 \pmod{p}$, whence p divides $a(x - y)$. Since $(a,p) = 1$, it follows that p divides $x - y$, i.e., $x \equiv y \pmod{p}$.

Besides solving congruences, systems of congruences can be solved, as the following ancient result will prove:

Theorem 2-2. *Chinese Remainder Theorem.* Let $m_1, \ldots, m_k$ be integers such that $(m_i,m_j) = 1$ for $i \neq j$. Then, the system

$$\begin{cases} x \equiv a_1 \pmod{m_1} \\ \cdot \\ \cdot \\ \cdot \\ x \equiv a_k \pmod{m_k} \end{cases} \tag{2-1}$$

where $a_1, \ldots, a_k$ are given integers, can be solved for an integer x. Moreover, if y is another integral solution of the congruences (2-1), then $y = x \pmod{m_1 \ldots m_k}$.

Proof: Unlike many "existence theorems" on solutions, the proof we give will actually show how a solution x to the system (2-1) is found in practice.

Let $M_i = (1/m_i)m_1 \ldots m_k$. Of course, M_i is an integer because the numerator $m_1 \ldots m_k$ contains the factor m_i. Because the moduli are pairwise relatively prime, $(m_i,M_i) = 1$. Now the congruence $M_iz \equiv 1 \pmod{m_i}$ can be solved for z (see Theorem 2-1). Let z_i denote one such solution and form the integer $x = M_1z_1a_1 + \ldots + M_kz_ka_k$. Then x will satisfy all the congruences in the system (2-1). For example, $M_2 \equiv \ldots \equiv M_k \equiv 0 \pmod{m_1}$ so $x \equiv M_1z_1a_1 + 0 \equiv a_1 \pmod{m_1}$. Similarly, $x \equiv a_i \pmod{m_i}$ for all i.

Now, if y is another solution to Equation (2-1), then $x \equiv y \pmod{m_1}, \ldots, x \equiv y \pmod{m_k}$, whence, $x - y$ is divisible by $m_1, \ldots, m_k$. But this implies $x - y$ is divisible by the product $m_1 \ldots m_k$ (see Exercise 1-1, Problem 5), or $x \equiv y \pmod{m_1 \ldots m_k}$.

Example 2-1. Since $(10,29) = 1$ we may solve

$$\begin{cases} x \equiv 1 \pmod{10} \\ x \equiv 3 \pmod{29} \end{cases} \tag{2-2}$$

First, we solve $29z \equiv 1 \pmod{10}$ and $10z \equiv 1 \pmod{29}$. Since $29 \equiv 9 \pmod{10}$, the first of these congruences is equivalent to $9z \equiv 1 \pmod{10}$, and by inspection a solution to that congruence is $z \equiv -1 \pmod{10}$. The congruence $10z \equiv 1 \pmod{29}$ has $z = 3$ as a solution. Therefore, a solution is $x = 29(-1)(1) + 10(3)(3) = 61$. To find all solutions we note that if y is any other solution, then $x \equiv y \pmod{10.29}$ or $x - y = 290t$ for some integer t. Thus, the general solution is $y = 61 - 290t$, where t is an arbitrary integer.

2-2 The Euler ϕ-Function

An important number theoretic function is the Euler ϕ-function. For a positive integer x, $\phi(x)$ is defined to be the number of positive integers $\leq x$ which are relatively prime to x. In this section a formula for $\phi(x)$ will be found by a process which is characteristic in number theoretic investigations. That is, the problem of finding such a formula is reduced to finding $\phi(x)$ when x is a prime power.

Lemma 2-1. If p is a prime and n is a positive integer, then $\phi(p^n) = p^n - p^{n-1}$.

Proof: The only integers in the sequence $1,2,\ldots,p^n$ which are multiples of p are $p,p^2,\ldots,p^{n-1}p$. Thus, this sequence contains exactly p^{n-1} multiples of p. Hence, $\phi(p^n) = p^n - p^{n-1}$.

Theorem 2-3. The ϕ-function is multiplicative in the sense that $\phi(ab) = \phi(a)\phi(b)$ for relatively prime integers a and b.

Proof: Let $r_1,\ldots,r_{\phi(b)}$ denote the positive integers $\leq b$ which are relatively prime to b and $s_1,\ldots,s_{\phi(a)}$ denote those which are $\leq a$ and relatively prime to a.

Consider the $\phi(a)\phi(b)$ integers $ar_i + bs_j$, $i = 1,\ldots,\phi(b)$, $j = 1,\ldots,\phi(a)$.

Observe that $(ar_i + bs_j, ab) = 1$. For if p is a prime dividing $ar_i + bs_j$ and ab, then in particular p divides a or b. If p divides a, then also, p divides bs_j, and as $(b,p) = 1$, p divides s_j. However, this contradicts the fact that $(s_j,a) = 1$. Equally, if p divides b we obtain a similar contradiction. Therefore, the integers $ar_i + bs_j$ are prime to ab.

Moreover, modulo ab the integers $ar_i + bs_j$ are all distinct. For suppose $ar_i + bs_j \equiv ar_k + bs_l \pmod{ab}$. Then, $a(r_i - r_k) - b(s_j - s_l)$ is a multiple of ab, and in particular a multiple of a. Therefore, a divides $b(s_j - s_l)$; whence, a divides $s_j - s_l$. However, this is possible only if $j = l$. Similarly, $a(r_i - r_k) - b(s_j - s_l)$ is a multiple of b and b divides $r_i - r_k$, which forces $i = k$.

Now if we reduce the $\phi(a)\phi(b)$ integers $ar_i + bs_j$ to their remainders by dividing them by ab, we will obtain $\phi(a)\phi(b)$ integers between 1 and ab which are relatively prime to ab. Thus, $\phi(ab) \geq \phi(a)\phi(b)$.

To show $\phi(ab) = \phi(a)\phi(b)$, it is sufficient to show that if c is any integer such that $(c,ab) = 1$ and $1 \leq c < ab$, then $c \equiv ar_i + bs_j \pmod{ab}$ for some i and j. However, for such an integer c the congruence $ax \equiv c \pmod{b}$ may be solved for x. Thus, $ax - c = tb$ for an integer t. Since $(c,ab) = 1$, the integer t must be relatively prime to a. Similarly, $(x,b) = 1$. Thus, $-t \equiv s_j \pmod{a}$ and $x \equiv r_i \pmod{b}$ for some i and j. Altogether, $c = ax - tb \equiv ar_i + bs_j \pmod{ab}$.

Corollary 2-1. Let $x = p_1^{e_1} \dots p_k^{e_k}$, where $p_1, \dots, p_k$ are distinct primes. Then $\phi(x) = (p_1^{e_1} - p_1^{e_1-1}) \dots (p_k^{e_k} - p_k^{e_k-1}) = x(1 - 1/p_1) \dots (1 - 1/p_k)$.

Proof: A repeated application of Theorem 2-3 shows $\phi(x) = \phi(p_1^{e_1}) \dots \phi(p_k^{e_k})$ and the preceding formulas follow from Lemma 2-1.

Theorem 2-4. *Euler's Theorem.* If $(a,n) = 1$, then $a^{\phi(n)} \equiv 1 \pmod{n}$.

Proof: Let $r_1, \dots, r_{\phi(n)}$ denote the integers between 1 and n which are relatively prime to n. Then the integers $ar_1, \dots, ar_{\phi(n)}$ are also relatively prime to n; hence, each of these integers is congruent modulo n to an integer in the set $r_1, \dots, r_{\phi(n)}$. Moreover, it is not difficult to see $ar_i \not\equiv ar_j \pmod{n}$ whenever $i \neq j$. Therefore, modulo n the integers $ar_1, \dots, ar_{\phi(n)}$ coincide, in some order, with the integers $r_1, \dots, r_{\phi(n)}$. Hence, taking products of the integers in these two set modulo n we have

$$a^{\phi(n)} r_1 \dots r_{\phi(n)} \equiv r_1 \dots r_{\phi(n)} \pmod{n}$$

Thus, $(a^{\phi(n)} - 1) r_1 \dots r_{\phi(n)}$ is divisible by n; whence, n divides $a^{\phi(n)} - 1$.

Theorem 2-5. *Fermat's Theorem.* If p is a prime and $a \not\equiv 0 \pmod{p}$ then $a^{p-1} \equiv 1 \pmod{p}$.

Proof: This follows immediately from Euler's Theorem and the fact that $\phi(p) = p - 1$.

With the aid of Euler's Theorem the congruence $ax \equiv b \pmod{n}$ can be easily solved if $(a,n) = 1$. Previously we solved such a congruence by turning to the Diophantine equation $ax - tn = b$. However, since $a^{\phi(n)} \equiv 1 \pmod{n}$, if the congruence is multiplied by $a^{\phi(n)-1}$ we obtain the solution $x \equiv a^{\phi(n)-1} b \pmod{n}$. For example, let us solve $7x \equiv 6 \pmod{48}$. Since $48 = 2^4 \cdot 3$, $\phi(48) = \phi(2^4)\phi(3) = (2^4 - 2^3) \cdot 2 = 16$, and $7^{16} \equiv 1 \pmod{48}$. Thus, $x \equiv 7^{15} \cdot 6 \equiv (7^2)^7(7)(6) \equiv 1^7 \cdot 42 \equiv 42 \pmod{48}$.

2-3 The Algebra of Equivalence Classes

We return to the theme that a congruence is rather like an equation. Our object will be to show in fact that a congruence can be regarded as an equation, but not as an equation involving a pair of integers. Instead, it will be an equation involving a pair of *sets* of integers.

The key here is the observation made at the beginning of our study that in congruence modulo n, integers are distinguished only by their remainders when divided by n.

Let S be a set or collection of objects and suppose there is given a relation $\backsim$ among certain pairs of the objects of S. Let us write $a \backsim b$ for objects a and b of S which are related via $\backsim$, and write $a \not\backsim b$ if a and b are not $\backsim$-related. The relation $\backsim$ is called an equivalence relation if the following are true:

(I) $a \backsim a$ for all a in S (reflexive property).
(II) $a \backsim b$ implies $b \backsim a$ (symmetry property).
(III) $a \backsim b$ and $b \backsim c$ imply $a \backsim c$ (transitive property).

Congruence modulo n clearly satisfies (I) and (II), while if $a \equiv b \pmod{n}$ and $b \equiv c \pmod{n}$, then $a - b = tn$ and $b - c = sn$ for integers s and t. Thus, $a - c = (a - b) + (b - c) = (t + s)n$, and $a \equiv c \pmod{n}$. Thus, congruence also obeys (III), which means that it defines an equivalence relation on the set of integers. (The astute reader will probably note that we have in fact already implicitly used the properties (I)–(III) in working with congruences.) An example of a relation which is not an equivalence relation is that of inequality ($\geq$) in the set of real numbers — it satisfies (I) and (III), but not (II).

Roughly speaking, when an equivalence relation is defined on a set, we look at these objects only from the point of view of the given relation. That is, distinct objects are not distinguished if they are equivalent with respect to the given relation.

With this in mind we define now $[a]$ to be the set of all objects x from the set S which are equivalent to a under the given equivalence relation $\backsim$. That is, $[a]$ consists of those x for which $a \backsim x$. Note that $[a]$ contains at least one element because of (I). We call $[a]$ the *equivalence class containing a*. The basic property of equivalence classes is given by the following proposition:

Proposition 2-2. Equivalence classes $[a]$ and $[b]$ are either identical or have no object in common.

In other words an equivalence relation partitions a set S into separate regions or subsets of equivalent objects.

To prove the proposition, let us suppose $[a]$ and $[b]$ are given equivalence classes which have an object u in common. This means $a \backsim u$ and $b \backsim u$. Our task is to show $[a]$ and $[b]$ consist of exactly the same objects. Since $\backsim$ is symmetric, $u \backsim b$. Thus, $a \backsim b$ because of transitivity; whence, b is a member of the class $[a]$. In fact, every object in $[b]$ is a member of $[a]$. For if v is in $[b]$, then $b \backsim v$, and as $a \backsim b$, $a \backsim v$ by transitivity again.

By a similar argument it is shown that a is in $[b]$ and every member of $[a]$ is a member of $[b]$. Thus, $[a] = [b]$ if a and b have at least one object in common.

We now calculate the equivalence classes which partition the integers under congruence modulo n.

Proposition 2-3. For a given positive integer n, congruence modulo n partitions the set of integers into exactly n distinct equivalence classes. These are $[0], [1], \ldots, [n - 1]$.

Proof: If r and r' are integers such that $0 \leq r < r' < n$, then $r' - r$ is not divisible by n, so $r' \not\equiv r \pmod{n}$. Thus, $[r']$ and $[r]$ are distinct equivalence classes. In particular, $[0], [1], \ldots, [n - 1]$ are all distinct.

Moreover, if a is any integer, then $a = qn + r$ for integers q and r, where $0 \leq r < n$. Then $a \equiv r \pmod{n}$ and a belongs to the class $[r]$. In other words, the equivalence classes $[r]$, $r = 0, \ldots, n - 1$ contain all the integers.

We now define the *sum* and *product* of equivalence classes $[a]$ and $[b]$ (modulo an integer n) as follows:

$$[a] + [b] = [a + b] \qquad \text{(addition rule)} \qquad \textbf{(2-3)}$$

and

$$[a][b] = [ab] \qquad \text{(product rule)} \qquad \textbf{(2-4)}$$

While the preceding rules are quite simple, a complication arises: For example, the addition of two equivalence classes is accomplished by choosing two integers from the respective classes, adding them, and then finding the equivalence class containing their sum. The problem is that $[a]$ contains many different integers. What would have happened had we chosen an integer in $[a]$, but different from a and then applied the addition rule? As a concrete example, let $n = 14$. Then $[2] + [3] = [5]$ by the addition rule, Equation (2-3). However, $[2] = [16]$ and $[3] = [31]$; so, if we apply Equation (2-3) using $a = 16$ and $b = 31$ instead of $a = 2$ and $b = 3$, we obtain $[2] + [3] = [47]$ rather than $[2] + [3] = [5]$. However, while $47 \neq 5$, $[47] = [5]$ because $47 \equiv 5 \pmod{14}$.

In general, the addition and product rules, Equations (2-3) and (2-4) respectively, are well-defined for arbitrary n in the sense that if we use different representatives from the pairs of equivalence classes we still get the same equivalence classes when forming their sum and product. Explicitly, if $[a] = [c]$ and $[b] = [d]$, then still $[c] + [d] = [a + b]$ and $[c][d] = [ab]$. For if $[a] = [c]$ and $[b] = [d]$, then $a \equiv c \pmod{n}$ and $b \equiv d \pmod{n}$, so, $a + b \equiv c + d \pmod{n}$, because of Proposition 2-1. Therefore, $[a + b] = [c + d]$. Similarly, $[ab] = [cd]$.

The necessity of showing that Equations (2-3) and (2-4) are *well-defined* rules is delicate and quite important. In later work, we shall encounter similar situations; so, we repeat in a general way what we have just done in the preceding paragraphs: We gave a rule for combining two sets of objects to get another set. This rule required that we pick a representative from each set, combine them, and then, find the set containing the result. What had to be shown was that the ultimate result was independent of the particular choice of representatives from the sets.

Returning now to congruences, it is easy to see that the congruence $ax \equiv b \pmod{n}$ is essentially the equation $[a]x = [b]$, where x denotes an equivalence class, rather than an integer.

Altogether, the system of equivalence classes has been given an arithmetic structure by the addition and multiplication rules, Equations (2-3) and (2-4). We shall denote this arithmetic system by $I/(n)$.

The sort of equation that can be solved in $I/(n)$, as well as the general arithmetic character of this system, depends very much on the prime divisors of the modulus n, as we shall see in the next section. However, for the moment let us examine two specific cases.

> $I/(6)$: The equation $[2]x = [4]$ has two distinct solutions, namely $x = [2]$ and $x = [5]$. The equation $[2]x = [1]$ has no solution x in this system, while $[5]x = [1]$ has exactly one solution, $x = [5]$. Thus, in $I/(6)$ an equation $[a]x = [b]$ can have several, none, or exactly one solution.
>
> $I/(5)$: The equation $[a]x = [b]$ can always be solved if $[a] \neq [0]$, because of Theorem 2-1. (Moreover, the solution is unique.) In this respect, $I/(5)$ behaves better than the integers themselves, for there is no integer x such that $3x = 2$ but in $I/(5)$ there is an equivalence class x for which $[3]x = [2]$.

2-4 Rings and Fields

In the previous section we constructed an algebraic system $I/(n)$ for $n = 1,2, \ldots$. In addition to these systems, we have the systems of integers, rational numbers, and real numbers, all of which are alike in some ways and different in others. To compare effectively these various algebraic structures it is convenient to think of them as examples of more general systems.

Let R be a set of elements in which there are two ways of combining a pair of elements a and b to get a third element of R. Call these operations "addition" and "multiplication" for convenience, and let $a + b$ and ab respectively denote the results of combining a and b by adding them and multiplying them. Then R is called a *ring* if the following properties hold for its elements and the two given operations:

(A1) For all a,b in R, $a + b$ is in R (closure law for addition).

(A2) For all a,b,c in R, $(a + b) + c = a + (b + c)$ (associative law for addition).

(A3) $a + b = b + a$ for all a,b in R (commutative law for addition).

(A4) R contains an element, which we denote by the symbol 0 with the property $0 + a = a$ for all a in R (law on the existence of an additive neutral element).

(A5) For each a in R there exists an element a' in R such that $a + a' = 0$ (subtraction axiom).

(M1) ab is in R for all a,b in R (closure law for multiplication).

(M2) For all a,b,c in R, $(ab)c = a(bc)$ (associative law for multiplication).

(M3) $a(b + c) = ab + ac$ and $(b + c)a = ba + ca$ for all a,b,c in R (distributive laws).

To give an example of a ring R we must specify the elements of R together with the two operations of combining them. Then we must verify the laws (A1)–(M3) for the system R.

Now, the set I of integers under the operations of ordinary addition and multiplication is a ring. Equally, the set Q of rational numbers and the set of real numbers form rings under ordinary addition and multiplication.

Another example of a ring is the set M_2 of all 2×2 matrices with the operations of matrix addition and multiplication. In this ring the "zero" element of axiom (A4) is $\begin{bmatrix} 0 & 0 \\ 0 & 0 \end{bmatrix}$. The ring M_2 is an example of a non-commutative ring. That is, if

$$a = \begin{bmatrix} 0 & 1 \\ 0 & 0 \end{bmatrix} \quad \text{and} \quad b = \begin{bmatrix} 1 & 0 \\ 0 & 0 \end{bmatrix}$$

then, $ab \neq ba$ since

$$ab = \begin{bmatrix} 0 & 0 \\ 0 & 0 \end{bmatrix} \quad \text{and} \quad ba = \begin{bmatrix} 0 & 1 \\ 0 & 0 \end{bmatrix}$$

From the fact that the integers form a ring and the addition and multiplication rules, Equations (2-3) and (2-4), it follows that the systems $I/(n)$ are rings, and the multiplication is commutative in these rings in the sense that $ab = ba$ for all a and b in $I/(n)$. The zero element of $I/(n)$ is the equivalence class [0].

The set of negative integers is not a ring because it violates (M1). Neither is the system of positive integers a ring — it does not satisfy (A5).

Before explaining the roles of the various ring axioms, let us develop a few elementary consequences from them.

Proposition 2-4. In a ring R the zero element of (A4) is unique, and the element a' of (A5) is uniquely determined by a. Moreover, $0a = a0 = 0$ for all a in R.

Proof: Let $0'$ be another element of R with the property $0' + a = a$ for all a in R. Then for the choice $a = 0$, we have $0' + 0 = 0$. On the other hand, from the defining property of the element 0, $0' + 0 = 0'$. Thus, $0' = 0$, i.e., the zero element is unique.

Given a, let a'' also have the property $a + a'' = 0$, and consider $a' + (a + a'')$. We have $a' + (a + a'') = a' + 0 = a'$. However, $a' + (a + a'') = (a' + a) + a'' = 0 + a'' = a''$. Therefore, $a'' = a'$, as required.

Now the fact that $0a = 0$ follows from one of the distributive laws. For $0a = (0 + 0)a = 0a + 0a$. Thus, $0a = 0a + 0a$. But, there exists an element $(0a)'$ such that $0a + (0a)' = 0$. Adding this element to both sides of the equation $0a = 0a + 0a$, we have $0 = 0a + (0a)' = (0a + 0a) + (0a)' = 0a + (0a + (0a)') = 0a + 0 = 0a$. Thus, $0 = 0a$. The proof of the other relation $a0 = 0$ is similar, but uses the other distributive law.

Since the element a' of axiom (A5) is uniquely determined by a, it is denoted by the more natural notation $-a$, rather than a', and $-a$ is called the *additive inverse* of a. In a ring R, subtraction may be introduced by defining for a,b in R, $a - b$ to be $a + (-b)$.

Roughly, in rings arithmetic involving addition may be carried out without too much difficulty. The axiom (M3) connects the two operations of addition and multiplication and permits factoring. However, as the multiplication is governed by fewer laws than the addition, computations involving products must be carried out with more care than those involving just addition alone. For example, the products ab and ba need not be the same, as the earlier matrix example showed.

A *field* F is a ring with at least two elements satisfying the further axioms:

(M4) $ab = ba$ for all a,b in F (commutative law for multiplication).
(M5) There exists an element 1 in F such that $1a = a$ for all a in F (existence law for a multiplicative neutral element).
(M6) For each $a \neq 0$ in F, there exists an element a' in F such that $aa' = 1$.

Entirely similar to Proposition 2-4 in content (and proof) is the following proposition:

Proposition 2-5. The element 1 of axiom (M5) is unique, as is the element a' on (M6).

The element 1 of a field is called the *identity* (sometimes, *unity*) *element*. The element a' of (M6) will be denoted by the more suggestive notation a^{-1} and it is called the *multiplicative inverse* of a.

We insist that a field F have at least two elements; otherwise, we would have the unhappy possibility that $1 = 0$ in F!

Examples of fields are the real numbers and Q, the rational numbers. However, the integers I do not form a field — for example, there is no x in I for which $2x = 1$. Thus, (M6) fails for the integers.

Another example of a field is the set $Q(\sqrt{2})$ of all real numbers of the form $x + y\sqrt{2}$, where x and y are rational numbers. To verify this, note that since the elements of $Q(\sqrt{2})$ are real numbers and the set of all real numbers is a field, the laws (A2), (A3), (M2), (M3), and M(4) automatically hold in $Q(\sqrt{2})$. Now, if x,x',y,y' are rational numbers, then $x + x'$ and $y + y'$ are rational. Hence, if $a = x + y\sqrt{2}$ and $b = x' + y'\sqrt{2}$ are arbitrary numbers of $Q(\sqrt{2})$, then so is $a + b = (x + x') + (y + y')\sqrt{2}$ in $Q(\sqrt{2})$. Therefore, (A1) is valid. Similarly, $ab = (x + y\sqrt{2})(x' + y'\sqrt{2}) = (xx' + 2yy') + (xy' + yx')\sqrt{2}$ is in $Q(\sqrt{2})$, which establishes (M4). Furthermore, $Q(\sqrt{2})$ contains the real numbers 0 and 1, as $0 = 0 + 0\sqrt{2}$

and $1 = 1 + 0\sqrt{2}$. Thus, (A4) and (M5) are true for $Q(\sqrt{2})$. Also, $-a = (-x) + (-y)\sqrt{2}$ is in $Q(\sqrt{2})$; hence, (A5). Finally, if $a = x + y\sqrt{2}$ is a nonzero element of $Q(\sqrt{2})$ then $Q(\sqrt{2})$ must contain a^{-1} because $(x + y\sqrt{2})^{-1} = 1/(x + y\sqrt{2}) = [1/(x + y\sqrt{2})][(x - y\sqrt{2})/(x - y\sqrt{2})] = (x/x^2 - 2y^2) + (-y/x^2 - 2y^2)\sqrt{2}$ and $x/x^2 - 2y^2$ and $-y/x^2 - 2y^2$ are rational. (But why is $x^2 - 2y^2 \neq 0$?) Thus, (M6) has been proved and $Q(\sqrt{2})$ is a field. Notice that $x + y\sqrt{2}$ already had a multiplicative inverse in the real field, but it had to be shown in the illustration of (M6) that its inverse was contained in the smaller $Q(\sqrt{2})$.

The field $Q(\sqrt{2})$ is actually larger than the rational field Q. For if x is any rational number, then $x = x + 0\sqrt{2}$, which shows x is contained in $Q(\sqrt{2})$. On the other hand, $Q(\sqrt{2})$ contains $\sqrt{2} = 0 + 1\sqrt{2}$, whereas, $\sqrt{2}$ is not in Q. In a later chapter, we shall see that in a sense $Q(\sqrt{2})$ is the smallest field containing Q and $\sqrt{2}$.

Among the rings $I/(n)$, we also find fields, and unlike the previous examples, these contain only finitely many elements.

Theorem 2-6. If p is a prime, then $I/(p)$ is a field with exactly p elements.

Proof: We have already noted that $I/(p)$ is a ring of p elements and that the commutative law for multiplication is valid in $I/(p)$. Moreover, $I/(p)$ contains an identity element, namely [1].

To verify (M6), let $[a]$ be a nonzero element of $I/(p)$. Then $a \not\equiv 0 \pmod{p}$; whence, the congruence $ax \equiv 1 \pmod{p}$ has a solution x. In terms of equivalence classes, $[a][x] = [1]$. Therefore, $[a]$ has $[x]$ as its multiplicative inverse.

For rings we have the following cancellation property:

Proposition 2-6. If a, b, and c are elements of a ring R and $a + c = b + c$, then $a = b$.

Proof: The c in the preceding equation is canceled out simply by adding $-c$ to both sides of the equation.

For fields we have the following analogous proposition:

Proposition 2-7. If a, b, and c are elements of a field F such that $ac = bc$ and c is nonzero, then $a = b$.

Corollary 2-2. If a and c are elements of a field such that $ac = 0$, then either $a = 0$ or $c = 0$.

Proof: If $c \neq 0$, then the equation $ac = 0$ may be written as $ac = 0c$; whence, $a = 0$ because of Proposition 2-7.

Sometimes Corollary 2-2 is stated as saying a field has no divisors of zero.

The fact that a field F has no divisors of zero proves useful in solving equations in F. For example, we may solve $0 = x^2 + [9]$ in the field $I/(13)$ by noting $[9] = [-4]$; whence, $0 = x^2 + [9]$ becomes $0 = x^2 - [4] = (x - [2])(x + [2])$. Since $I/(13)$ has no divisors of zero, it follows that either $x = [2]$ or $x = -[2]$.

It is possible to sharpen Theorem 2-6 to get the following theorem:

Theorem 2-7. The ring $I/(n)$ is a field if and only if n is a prime.

Proof: Suppose n is not prime, then it can be factored $n = ab$ as a product of smaller integers a and b. Of course, $a \not\equiv 0 \pmod{n}$ and $b \not\equiv 0 \pmod{n}$. So, $[a] \neq [0]$ and $[b] \neq [0]$. However, $[a][b] = [n] = [0]$. Since we have exhibited a pair of zero divisors in $I/(n)$, it follows that $I/(n)$ is not a field.

In case a ring or field has only finitely many elements, it is sometimes convenient to display the addition and multiplication rules by tables. To accomplish this we list in fixed order the elements of the ring along the top and side of the table. Then to find for example the product ab of elements a and b, we look in the row indexed by a and the column indexed by b. The cell at which the row intersects the column will contain the entry ab. For example, for the field $I/(3)$, whose elements are [0], [1], and [2], Tables 2-1 and 2-2 are constructed.

Table 2-1

+	[0]	[1]	[2]
[0]	[0]	[1]	[2]
[1]	[1]	[2]	[0]
[2]	[2]	[0]	[1]

Table 2-2

×	[0]	[1]	[2]
[0]	[0]	[0]	[0]
[1]	[0]	[1]	[2]
[2]	[0]	[2]	[1]

2-5 Analytic Geometry

Roughly speaking, a field is a system in which we may add, subtract, multiply, and divide just as we do for real numbers. Of course, there are fundamental differences between $I/(5)$ and the real numbers for example, but these differences do not occur so much in the arithmetic. And now, rather than describing or classifying fields according to their algebraic properties, we shall show how a well-known construction from real numbers can be repeated for arbitrary fields. What we have in mind is the application of algebra to geometry, i.e., Descartes' analytic geometry.

In ordinary planar geometry the following are true:

(i) Through any pair of distinct points there passes exactly one (straight) line.
(ii) Two distinct lines are either parallel or intersect in just one point.
(iii) Given a point P and a line l not passing through P, there exists exactly one line l' passing through P which is parallel to l.

The solution of geometric problems about points and lines is most simply accomplished by assigning coordinates (x,y) to a point P in the plane and describing the lines as graphs of algebraic equations. The vertical lines have equation $x = a$, while the lines of finite slope have equations of the type $y = mx + b$.

Now the problem of finding the lines guaranteed in (i) or (iii) is reduced to solving simple algebraic equations. Equally the point at which a pair of lines intersect is found in much the same way.

For example, the point $P = (1,2)$ does not lie on the line l whose equation is $y = 3x + 1$. The line l' of (iii) must have a slope equal to 3; hence, an equation of the form $y = 3x + b$. Since l' passes through P, $2 = 3\cdot 1 + b$; hence $b = -1$. Thus, l' is the line $y = 3x - 1$.

The equation we solved can be done in any field. Equally, in the other geometric problems, the equations encountered are of a type that can be handled in any field. In other words, the fact that we used the real number field, instead of for instance $I/(5)$ is not really essential in solving the geometric problems. This suggests that a geometric system similar to ordinary lines and points could be coordinatized by an arbitrary field. To allow for these new geometries, we must abstract the notion of "plane geometry" in much the same way we extended the idea of real number field to fields in general.

An *affine geometry* is a system Π of objects called "points" and collections of these points called "lines" such that the following are true:

(I) Any pair of distinct points is contained in exactly one line.
(II) Any pair of distinct lines either have one or no point in common.
(III) Given a point P and a line l which does not contain P, there exists exactly one line l' which contains P but does not contain any point in common with l.
(IV) The system Π contains at least two lines.

The purpose of axiom (IV) is to rule out rather uninteresting examples of affine geometries.

The ordinary Euclidean plane is an example of an affine geometry. For another, quite different, example let the points of Π be the letters A, B, C, and D. Each pair of these points should be joined by a unique line, and if for simplicity we suppose each line should contain no more than two points; then, we obtain $\binom{4}{2} = 4!/2!2! = 6$ lines, namely, $l_1 = \{A,B\}$, $l_2 = \{C,D\}$, $l_3 = \{A,D\}$, $l_4 = \{B,C\}$, $l_5 = \{B,D\}$, $l_6 = \{A,C\}$. (The notation $l_1 = \{A,B\}$ simply means the line l_1 consists of the points A and B.)

It is a straightforward matter to verify the system Π consisting of the points $A - D$ and the lines $l_1 - l_6$ satisfying (I)–(IV). Of course this affine geometry is radically different from the ordinary plane in that it consists of only 4 points and 6 lines. However, it does share with the plane the properties (I)–(IV).

A more reliable way of finding affine geometries is to coordinatize them by fields, as follows: Given a field F we form a geometry Π by letting the points of Π be the ordered pairs (x,y), where x and y belong to F. The lines of Π will be of two types. There are the lines $y = mx + b$ of slope m and the "vertical" lines $x = a$. Specifically, the line $y = mx + b$ consists of the points $(x,mx + b)$, where m and b are fixed elements of F and x takes on all values from F. Similarly, the line $x = a$ consists of those points (x,a), where a is fixed and x is not.

Let us carry out this program for $F = I/(2)$. This field contains the two elements [0] and [1], and the corresponding affine geometry consists of the points $A = ([0], [0])$; $B = ([0], [1])$; $C = ([1], [0])$; and $D = ([1], [1])$. There are two "vertical" lines $x = a$, depending on whether $a = [0]$ or $[1]$. The choice $a = [0]$ yields the line l_1 whose points are A and B. Thus, $l_1 = \{A,B\}$. For $a = [1]$ we obtain the line $l_2 = \{C,D\}$. For the lines $y = mx + b$ there are two choices for the value of m and two for that of b. Thus, we shall obtain four new lines of this general type. In fact, these are

$$l_3 = \{A,D\} \quad \text{(the line } y = x\text{)}$$
$$l_4 = \{B,C\} \quad \text{(the line } y = x + [1]\text{)}$$
$$l_5 = \{B,D\} \quad \text{(the line } y = [1]\text{)}$$
$$l_6 = \{A,C\} \quad \text{(the line } y = [0]\text{)}$$

Checking back, it will be seen that this example is identical to the previous one.

We have indicated how to construct geometries from fields. It is a remarkable and deep result that the converse is in a sense true. That is, given an abstract geometry satisfying (I)–(IV), coordinates may be introduced in such a way that the lines of the geometry are defined by equations in the algebraic system which produces the coordinates. The underlying algebraic object need not be a field; however, it does possess some of the field properties. Very often geometric properties of affine geometries can be distinguished according to the algebraic properties of the system which coordinatizes the geometry. Indeed, this has been one of the central areas of study in modern geometry.

2-6 Latin Squares

Finite fields are useful in constructing Latin squares, which in turn are significant in the design of stastical experiments and geometry.

A *Latin square* of order n is an n by n square of n^2 cells such that each of a given set of n symbols occurs once in each row and column of the square. For convenience we take the symbols to be the integers $0,1,\ldots,n-1$. An example of a 3 by 3 Latin square is Table 2-3.

Table 2-3

0	2	1
1	0	2
2	1	0

While Table 2-4 is an example of a 4 by 4 square.

Table 2-4

0	1	2	3
1	2	3	0
3	0	1	2
2	3	0	1

To see how Latin squares arise in the design of experiments let us imagine we wanted to test four brands of corn fertilizers, denoting each of these brands by 0, 1, 2, and 3. Now an obvious way of testing these fertilizers would be to use them on a plot of corn as shown in Table 2-5.

Table 2-5

brand 0
brand 1
brand 2
brand 3

However, the problem with placing the fertilizer according to Table 2-5 is that unseen conditions, such as soil or drainage, may affect the outcome of the experiment. Therefore, to minimize the effect of such conditions, it is better to disperse the fertilizers by subdividing the plot into 16 subregions. Then apply the fertilizers to these smaller plots in such a way that a given fertilizer is used once and only once in each row and column of the plot. An example of such a placement is the Latin square Table 2-6.

Table 2-6

0	1	2	3
1	2	3	0
3	0	1	2
2	3	0	1

To achieve even better statistical tests, orthogonal sets of Latin squares are quite useful. We shall not discuss how they are used, but we shall give a construction for them.

Let L_1 and L_2 be n by n Latin squares and denote the entries of row i, column j of L_1 and L_2 by a_{ij} and b_{ij} respectively. Then L_1 and L_2 are called *orthogonal* if the ordered pairs (a_{ij}, b_{ij}) are all distinct for $i,j = 1, \ldots, n$. In other words, if L_2 is superimposed on L_1 and the entries in each of the n^2 cells are written down in order, then the resulting pairs are all different. For example, see Tables 2-7 and 2-8.

Table 2-7

$L_1 =$

0	2	1
1	0	2
2	1	0

Table 2-8

$L_2 =$

0	2	1
2	1	0
1	0	2

Tables 2-7 and 2-8 are orthogonal because placing L_2 upon L_1 yields the array Table 2-9 in which all 3^2 entries are distinct.

Table 2-9

(0,0)	(2,2)	(1,1)
(1,2)	(0,1)	(2,0)
(2,1)	(1,0)	(0,2)

More generally, a set of Latin squares $L_1, L_2, \ldots$ is called an orthogonal set if each pair of distinct squares is an orthogonal set. Now, two questions arise regarding orthogonal sets of Latin squares: What is the maximum number of n by n squares in an orthogonal set? How can a maximal set be constructed?

Theorem 2-8. If $L_1, \ldots, L_t$ is an orthogonal set of n by n Latin squares, then $t \leq n$ 1.

Proof: Relabeling if necessary, we may make the first row of each of the squares to be

0	1	. . .	$n-1$

without destroying the orthogonality of the set. In comparing any two squares, it follows that the ordered pairs (0,0), (1,1), . . . , $(n-1, n-1)$ are used up in the first row. Therefore, in the shaded cell

0	1	. . .	$n-1$
(shaded)			

distinct squares must have distinct entries. Moreover, no entry of this cell can $= 0$ because the squares are Latin. Therefore, there are at most $n - 1$ orthogonal squares in the set.

Unfortunately the answer to the second question is not yet known. However, for a prime p it is possible to construct a maximal set of $p - 1$ orthogonal p by p squares $L_1, \ldots, L_{p-1}$. To form L_k one places in row i, column j the integer $(i - 1)k + (j - 1)$, and then reduces all the entries modulo p (i.e., the entries of L_k are the elements of the field $I/(p)$). Thus, the scheme for L_k is Table 2-10.

Table 2-10

0	1	$\cdot\ \cdot\ \cdot$	$p-1$
k	$k+1$	$\cdot\ \cdot\ \cdot$	$k+(p-1)$
$2k$	$2k+1$	$\cdot\ \cdot\ \cdot$	$2k+(p-1)$
$\cdot$ $\cdot$ $\cdot$	$\cdot$ $\cdot$ $\cdot$	$\cdot$ $\cdot$ $\cdot$	$\cdot$ $\cdot$ $\cdot$
$(p-1)k$	$(p-1)k+1$	$\cdot\ \cdot\ \cdot$	$(p-1)k$ $+(p-1)$

Carrying out this procedure for $p = 3$, we obtain Tables 2-11 and 2-12.

Table 2-11

$L_1 =$

0	1	2
1	2	0
2	0	1

Table 2-12

$L_2 =$

0	1	2
2	0	1
1	2	0

2-7 On Identifying Algebraic Systems

It is generally accepted that for the study of algebraic systems one is not really interested in what the elements of such a system are. Rather, he is only concerned with the way in which these elements are combined, be it called addition, multiplication, subtraction, etc.

This attitude is not a recent development in mathematics. Indeed, Euclid gave only the most cursory description of points and lines (eg., calling a point "that which has no part") and then plunged on to give volumes of deep and beautiful theorems about the figures constructed from points and lines.

If we accept the idea that the objects of an algebraic system are in themselves unimportant and that it is only the algebraic relations between the objects that really count, then this raises the following question: When are two algebraic systems identical or different? Clearly, we may not differentiate two such systems just by comparing their individual elements.

What we shall do is call systems S and T identical if

(I) S and T have the same number of elements.
(II) the algebraic relations between the elements of S are identical to the algebraic relations between the elements of T.

It is the purpose of this section to give precise meaning to (I) and (II).

A *mapping* f is a rule of correspondence which assigns to each element of a set S a unique element of a set T. (See Figure 2-1.)

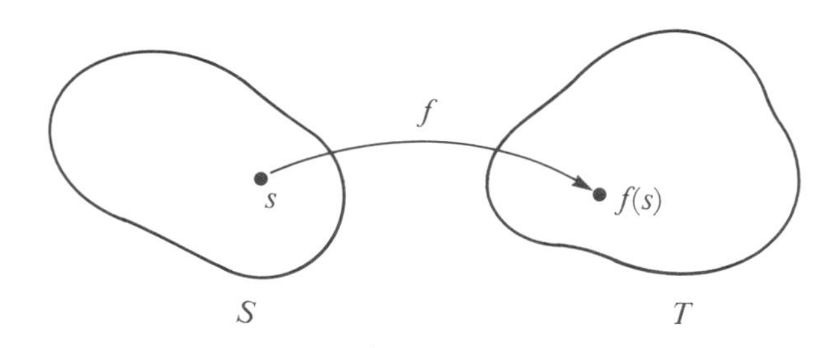

Figure 2-1

For an element of s of S, $f(s)$ denotes the unique element of T which corresponds to s and $f(s)$ is called the image of s. A mapping is just a function with domain S and range T. Often we write $f : S \to T$ to indicate that f is a mapping from S to T.

Now, it may happen that $s \neq s'$, but $f(s) = f(s')$. (See Figure 2-2.)

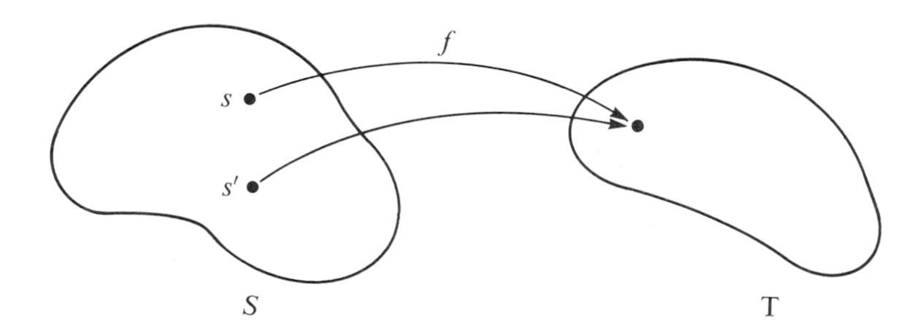

Figure 2-2

For example, the mapping $f: R \to R$ defined by $f(s) = s^2$, where R is the real field, has the property that $f(1) = f(-1)$. A mapping $f : S \to T$ is said to be *one-to-one* if $s \neq s'$ implies $f(s) \neq f(s')$ for s, s' in S. Thus, the preceding example is not one-to-one.

It can also be the case for a mapping $f : S \to T$ that not every element of T is of the form $f(s)$ for some s in S. In the preceding example, there is no s for which $f(s) = -3$. A mapping $f : S \to T$ is said to be *onto* if each element of T *is* of the form $f(s)$ for some s in S.

Equality between the number of elements of one set and another is described via mappings, even if the sets have infinitely many elements.

Roughly we say S and T have the same number of elements if the elements of S can be paired off with those of T. (See Figure 2-3.)

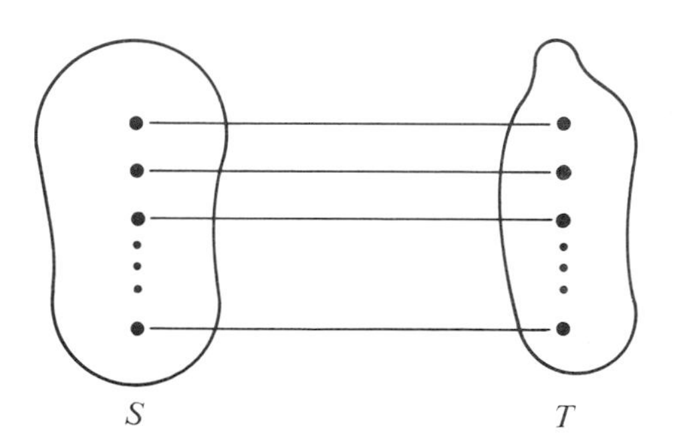

Figure 2-3

Precisely, we say a set S is *equipotent* to a set T (i.e., it has the same number of elements as T) if there exists a one-to-one onto mapping $f : S \to T$.

This definition is quite reasonable if S has finitely many elements. For if the elements of S are $s_1, \ldots, s_n$ and if S is equipotent to T, then there is a one-to-one onto map $f : S \to T$. The elements $f(s_1), \ldots, f(s_n)$ are all distinct because f is one-to-one; and these are all the elements of T because f is onto. Thus, S and T each have n elements.

Notice that according to this definition the set S of all integers is equipotent to the set T of just the even integers, because the mapping $f : S \to T$ defined by the rule $f(n) = 2n$ for an integer n is both one-to-one and onto.

Suppose S and T are rings and that S is equipotent to T, i.e., there exists a mapping $f : S \to T$ which is both one-to-one and onto. Suppose further that the addition and multiplication in the ring S are denoted by $+$ and $\times$ respectively, while $\oplus$ and $\otimes$ denote the addition and multiplication in T. If the mapping f has the further properties

$$f(a + b) = f(a) \oplus f(b) \quad \text{for all } a,b \text{ in } S \tag{2-5}$$

$$f(a \times b) = f(a) \otimes f(b) \quad \text{for all } a,b \text{ in } S \tag{2-6}$$

then the ring S is said to be *isomorphic* to T. Evidently the properties (2-5) and (2-6) show that the mapping f matches up the addition and multiplication operations of the ring S with those of the ring T. It is in this precise sense that the property (II) is expressed. The mapping f itself is called an *isomorphism*.

Altogether we shall regard the ring S to be algebraically identical to the ring T if there exists an isomorphism $f: S \to T$. Notationally it becomes awkward at times to use different symbols to distinguish the addition and multiplication of the ring S from the analogous operations of the ring T. Consequently, for an isomorphism $f: S \to T$ we will simply write

$$f(a + b) = f(a) + f(b) \tag{2-7}$$

$$f(ab) = f(a)f(b) \tag{2-8}$$

instead of the previous Equations (2-5) and (2-6). It must be understood, however, that the addition and multiplication on the left-hand and right-hand sides of Equations (2-7) and (2-8) take place in different rings.

If the ring S is isomorphic to the ring T, then by just a change of notation the addition and multiplication tables for T may be made identical to those for S. For example, if Table 2-13 is the addition table

Table 2-13

	. . .	b
. . .		
a		$a + b$
. . .	. . .	. . .

for S and Table 2-14 is the addition table for T, then because of Equation (2-7), the latter table becomes Table 2-15.

Table 2-14

	$\cdots$	$f(b)$
$\vdots$		
$f(a)$		$f(a) + f(b)$
$\vdots$	$\cdots$	$\vdots$

Now, by simply deleting the symbol f once from each cell of the last table, we will obtain the addition table for S.

If it is by the notion of isomorphism that we compare rings, then at the very least isomorphism should be an equivalence relation for rings. To show this is indeed the case, we require more general knowledge about mappings.

For mappings f and g (not necessarily isomorphisms) their *product* gf is defined by the rule

$$gf(s) = g(f(s)) \qquad s \in S \tag{2-9}$$

where $f: S \to T$, $g: T \to U$, and S, T, and U are sets (not necessarily rings). The product gf is a mapping $gf: S \to U$, and to map an element s of S via gf, we first apply the mapping f to s and then apply the mapping g to the resulting element $f(s)$. (See Figure 2-4.)

Table 2-15

	. . .	$f(b)$
. . .		
$f(a)$		$f(a+b)$
. . .	. . .	. . .

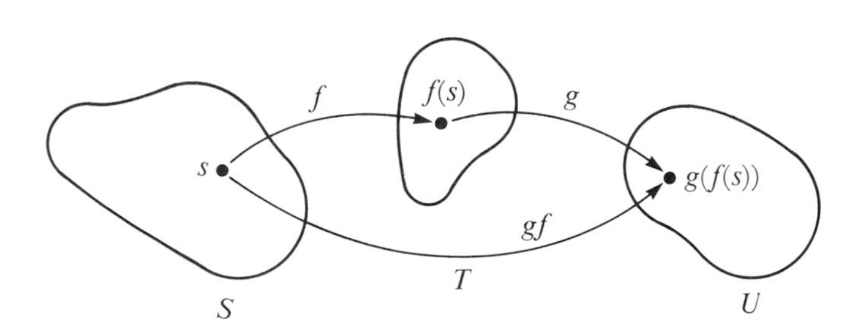

Figure 2-4

Having introduced a product for mappings, let us point out that there is an identity element for this product, though this identity mapping, which we indicate by I_S, depends on the set S. The mapping $I_S : S \to S$ is defined by the rule $I_S(s) = s$ for all s in S. If $f : S \to T$, then $f = fI_S$ in the sense that $f(s) = fI_S(s)$ for each s in S. Similarly, if $g : T \to S$, then $g = I_S g$.

Theorem 2-9. Let $f : S \to T$ be a mapping. Then there exists a mapping $g : T \to S$ such that $fg = I_T$ and $gf = I_S$ if and only if f is both one-to-one and onto. Moreover, the mapping g, if it exists, is also one-to-one and onto.

Proof: In the case that f is one-to-one and onto, we define the mapping $g : T \to S$ in the following way: Each element t of T is of the form $t = f(s)$ for some s in S (because f is onto). Moreover, there is only one element s of S for which $t = f(s)$, because f is one-to-one. We then define $g(t) = s$:

$$g(t) = s \quad \text{if and only if} \quad t = f(s) \tag{2-10}$$

It is immediate from Equation (2-10) that

$$fg = I_T \quad \text{and} \quad gf = I_S \tag{2-11}$$

Moreover, g is one-to-one. If $g(t) = g(t')$ for t, t' in T; then, $f(g(t)) = f(g(t'))$; so, $fg(t) = fg(t')$. But, $fg = I_T$; hence, $t = t'$. Also, g is onto since if s is an arbitrary element of S; then, $s = I_S(s) = gf(s) = g(f(s))$, i.e., s is the image of $f(s)$ under the mapping g.

To prove the converse, suppose there exists a mapping $g : T \to S$ satisfying Equation (2-11). Then for any t in T, $t = fg(t) = f(g(t))$; so, t is the image under f of the element $g(t)$. Thus, f is onto. Furthermore, if $f(s) = f(s')$ for s, s' in S, then $g(f(s)) = g(f(s'))$ or $gf(s) = gf(s')$. But, $gf = I_S$; hence, $s = s'$. Therefore, f is one-to-one, and this completes the proof of Theorem 2-9.

Remark 2-1. We wish to emphasize here that Theorem 2-9 is valid for arbitrary sets S and T. The mapping g of this theorem is unique in the sense that if also $fg_1 = I_T$ and $g_1 f = I_S$ for some other mapping $g_1 : T \to S$, then the mappings g and g_1 are identical, i.e., $g(t) = g_1(t)$ for all t in T. The mapping g is called the *inverse* of f and is denoted by $g = f^{-1}$.

With this fundamental result on mappings we are now prepared to prove the following theorem:

Theorem 2-10. Isomorphism is an equivalence relation among rings.

Proof: (i) *Reflexive Property.* We have to show each ring S is isomorphic to itself, or establish an isomorphism $f: S \to S$. This is done by simply taking $f = I_S$. (ii) *Symmetry.* If S is a ring which is isomorphic to a ring T, then there is an isomorphism $f: S \to T$. Since f is one-to-one and onto, there exists by the previous theorem an inverse mapping $f^{-1}: T \to S$. To prove T is isomorphic to S it is sufficient to verify Equations (2-7) and (2-8) for f^{-1} since f^{-1} is already one-to-one and onto, by virtue of Theorem 2-9.

For elements a and b of T, we have $f(f^{-1}(a + b)) = a + b$. Also, $a + b = f(f^{-1}(a)) + f(f^{-1}(b)) = f(f^{-1}(a) + f^{-1}(b))$ because f is an isomorphism. Thus, $f(f^{-1}(a + b)) = f(f^{-1}(a) + f^{-1}(b))$, which in turn implies $f^{-1}(a + b) = f^{-1}(a) + f^{-1}(b)$ because f is one-to-one. This shows Equation (2-7); the proof of Equation (2-8) is similar. (iii) *Transitivity.* If S, T, and U are rings such that S is isomorphic to T and T is isomorphic to U; then, isomorphisms $f: S \to T$ and $g: T \to U$ are given. The product gf of these isomorphisms may be verified to be an isomorphism $gf: S \to U$.

Exercises

Exercise 2-1

1. Solve the following congruences: $3x \equiv 5 \pmod{8}$, $6x \equiv 2 \pmod{16}$, $71x \equiv 3 \pmod{13}$.
2. Solve $x^4 \equiv 1 \pmod{5}$, $x^4 \equiv 1 \pmod{8}$.
3. Find the remainder when $1^3 + 2^3 + \ldots + 73^3$ is divided by 24.
4. Show that if p is a prime and r is an integer $0 < r < p$, then $\binom{p}{r} \equiv 0 \pmod{p}$.

   Here $\binom{n}{r} = n!/r!(n - r)!$. Is it true for all n that $\binom{n}{r} \equiv 0 \pmod{n}$?
5. Find all integers that give remainders 2, 3, and 4 when divided by 7, 8, and 25 respectively.
6. Show the system

$$\begin{cases} x \equiv a \pmod{n} \\ x \equiv b \pmod{m} \end{cases}$$

   has a solution if $a - b$ is divisible by (m,n).

Exercise 2-2

1. Find all integers x such that $\phi(x) = 8$.

2. If n is a positive integer, then the sum of the positive integers $\leq n$ which are relatively prime to n is $(n/2)\phi(n)$. Verify this for n a prime, a prime power, and then for arbitrary n.

3. Show the number of fractions $a/b < 1$, where $(a,b) = 1$ and $b \leq n$ is $\phi(1) + \ldots + \phi(n)$. Note: $\phi(1) = 1$.

4. Using induction on the integer n prove the following: If a is an odd integer and $n \geq 3$, then $a^{2^{n-2}} \equiv 1 \pmod{2^n}$.

Exercise 2-3

1. Suppose a set is partitioned into subsets in such a way that any two distinct subsets have no element in common. Define an equivalence relation on that set such that the given subsets are exactly the equivalence classes produced by the equivalence relation. (This shows that equivalence relations and partitions are essentially the same things, i.e., equivalent!)

2. Give examples of relations which satisfy two of the requirements for equivalence relation but not all the requirements.

3. In what sense does a tax table define an equivalence relation among taxpayers?

4. What is incorrect about the following argument: The property (I) is really a consequence of (II) and (III), for $a \sim b$ and $b \sim a$ by symmetry; hence, $a \sim a$ because of transitivity.

5. For equivalence classes modulo n define a composition * by the rule $[a]*[b] = [c]$, where $c =$ the larger of a and b. Is this rule well-defined?

6. Show that if p is a prime and $[a]$ and $[b]$ are elements of $I/(p)$, then $([a]+[b])^p = [a]^p + [b]^p$.

Exercise 2-4

1. Show that in rings, $-(-a) = a$, $-(a+b) = -a-b$, and $(-a)b = a(-b) = -ab$.

2. Do the real numbers of the form $a + b\sqrt[3]{2}$, where a and b are rational numbers, form a field?

3. In $I/(11)$, find the multiplicative inverses of [2] and [5].

4. In the ring $I/(8)$, find all solutions to $(x - [1])(x + [1]) = [0]$. Explain why there are solutions different from $x = [1]$ and $x = -[1]$.

5. Write down the addition and multiplication tables for the field $I/(5)$, for the ring $I/(6)$.

6. How are the cancellation and commutative laws reflected in the addition and multiplication tables for a field?

7. Let $F = I/(3)$ and $F(\theta)$ denote the set of all expressions $x + y\theta$, where a and y are in F, and products and sums are defined by

$$(x + y\theta)(x' + y'\theta) = (xx' + 2yy') + (xy' + yx')\theta$$
$$(x + y\theta) + (x' + y'\theta) = (x + x') + (y + y')\theta$$

Find the zero and identity elements of F. Verify that if $x + y\theta \neq 0$, then $x + y\theta$ has a multiplicative inverse. In fact, $F(\theta)$ turns out to be a field, much like the example $Q(\sqrt{2})$ in the text. The element θ is in a sense a "square root of 2." Show that $F(\theta)$ is actually "larger" than F. How many elements are in $F(\theta)$?

Exercise 2-5

1. Show there does not exist an affine geometry with exactly 3 points.
2. Show that the notion of parallelism is an equivalence relation on the lines of an affine geometry. (You will have to use the axioms to do this.)
3. Construct the affine geometry coordinatized by $I/(3)$.
4. How many points and lines are there in the affine geometry coordinatized by a field of n elements?
5. What goes wrong with the geometry when one attempts to coordinatize it with the ring $I/(4)$?
6. Let Π be the affine geometry coordinatized by the rational field Q and let C denote the unit circle $x^2 + y^2 = 1$. Where do the lines $y = mx$ intersect C? Answer the same question but with Π coordinatized by $I/(3)$, rather than Q.
7. Let Π be the geometry coordinatized by the field $F(\theta)$ of Exercise 2-4, Problem 7. Find the equation of the line l passing through the points $([1] + \theta, [2] + \theta)$ and $([2], [1])$. What are the other points of l?

Exercise 2-6

1. Show each of the squares L_k is a Latin square.
2. Verify the set $L_1, \ldots, L_{p-1}$ is orthogonal.
3. Explain how an orthogonal set of $n - 1$ Latin n by n squares could be constructed from a field of n elements.
4. Construct an orthogonal set of four 5 by 5 Latin squares.
5. The Andorran chessmaster Don José Maria Puig wishes to test 5 new defenses he has invented for playing the black pieces. In order to test the general soundness of these defenses, he plans to use them in the coming chess season in Andorra, which consists of 5 tournaments in which he and the next 5 strongest Andorran masters are the only participants. In each of these tournaments he will play one game with the black pieces against each

opponent. Don José's opponents have radically different styles and his own play can be erratic from one tournament to another. In view of this, decide which defenses should be employed against his opponents in each of the 5 tournaments in order to properly test the defenses.

Exercise 2-7

1. For the real field R which of the following mappings $f: R \to R$ are one-to-one? onto?
   (a) $f(x) = x^3 + 2x + 1$
   (b) $f(x) = [x]$ ($[x]$ = the greatest integer $\leq x$)
   (c) $f(x) = e^x \sin x$.
2. Complete the proof of part (iii) of Theorem 2-10.
3. Let $f: S \to T$ be an isomorphism of rings S and T.
   (a) Show that if S contains an identity element 1, then $f(1)$ is the identity of T.
   (b) Show that $f(0)$ is the zero element of T.
   (c) Show that if S is commutative, so is T.
4. Call a ring S *quasi-isomorphic* to a ring T if there exists a pair of mappings $f: S \to T$, $g: S \to T$ such that f is one-to-one and onto and $g(a+b) = g(a) + g(b)$, $g(ab) = g(a)g(b)$ for all a,b in S. Are quasi-isomorphic rings isomorphic? Hint: Try S = the ring of integers and T = the ring of even integers. Is quasi-isomorphism an equivalence relation for rings? What is wrong with considering S and T to be algebraically identical if they are quasi-isomorphic?
5. Let F be the ring $F - \{a,b,c\}$ with addition and multiplication Tables 2-16 and 2-17.

Table 2-16

+	a	b	c
a	a	b	c
b	b	c	a
c	c	a	b

Table 2-17

×	a	b	c
a	a	a	a
b	a	c	b
c	a	b	c

Find an isomorphism $f: I/(3) \to F$.

6. An isomorphism f between two affine geometries Π and Π' should be a one-to-one onto mapping between the points of Π and Π' which preserves geometric properties.
   (a) What should f do to the *lines* of Π?
   (b) Answering (a), arrive at a simple definition of isomorphism for affine geometries.
   (c) Let Π' be the affine geometry consisting of the points W,X,Y,Z and the lines $l_1' = \{X,Z\}$, $l_2' = \{W,Y\}$, $l_3' = \{W,Z\}$ $l_4' = \{X,Y\}$, $l_5' = \{Y,Z\}$, $l_6' = \{W,X)$. Show that Π' is isomorphic to the affine geometry in Section 2-5.

3

Polynomials and Integral Domains

3-1 The Arithmetic of Polynomials

In calculus the simplest functions studied are the polynomials, i.e., expressions of the form

$$f(x) = a_0 + a_1x + \ldots + a_nx^n \tag{3-1}$$

where the coefficients $a_0,a_1, \ldots, a_n$ of $f(x)$ are real numbers and x is a variable. The purpose of this section is to point out that many operations involving polynomials may be carried out even when the coefficients belong to an abitrary field.

Let F be a field. An expression of the form of Equation (3-1) where the coefficients $a_0,a_1, \ldots, a_n$ belong to F is called a *polynomial*. The totality of all such polynomials is denoted by $F[x]$. The *degree* of a polynomial $f(x)$ is defined to be the highest power of x actually present in $f(x)$. Thus, in Equation (3-1) if $a_n \neq 0$, then $f(x)$ has degree n. The degree of $f(x)$ will often be denoted as simply $\deg f(x)$.

Polynomials $f(x) = a_0 + a_1x + \ldots + a_nx^n$ and $g(x) = b_0 + b_1x + \ldots + b_mx^m$ are called *equal* if their respective coefficients are equal, i.e., if $a_0 = b_0, a_1 = b_1, \ldots$.

Addition and multiplication of polynomials are done in a natural way. That is, if $f(x) = a_0 + a_1x + \ldots + a_nx^n$ and $g(x) = b_0 + b_1x + \ldots + b_mx^m$, then the *sum* $f + g$ and *product* fg of these polynomials are defined by

$$f(x) + g(x) = (a_0 + b_0) + (a_1 + b_1)x + \ldots \qquad \textbf{(3-2)}$$

and

$$f(x)g(x) = a_0b_0 + (a_1b_0 + a_0b_1)x + (a_2b_0 + a_1b_1 + a_0b_2)x^2 + \ldots \qquad \textbf{(3-3)}$$

(From a logical point of view it is perhaps disturbing to define addition and multiplication of polynomials when in the definition [Equation (3-1)] of a polynomial there already appears addition among the various terms $a_0, a_1x, \ldots$. However, this logical difficulty is easily removed — see Exercise 3-2, Problem 2.)

The equality rule, together with Equations (3-2) and (3-3) permit one to operate (algebraically) with polynomials over an arbitrary field F in the same way one handles polynomials with real coefficients. In fact, we have the following proposition:

Proposition 3-1. If F is a field, then $F[x]$ is a commutative ring with an identity element.

The proof of this result is straightforward and will be omitted. We only remark that the zero element of $F[x]$ is the polynomial, all of whose coefficients $= 0$, and the identity element is the constant polynomial $f(x) = 1$. The ring $F[x]$ is called the *ring of polynomials over* F.

The polynomial ring $F[x]$ is remarkably similar to the ring of integers, and this similarity will have profound applications to the theory of fields. The key to this analogy is the following division algorithm:

Theorem 3-1. Let $f(x)$ and $g(x)$ be polynomials of $F[x]$, where F is a field, and suppose $g(x)$ is not the zero polynomial. Then there exist polynomials $Q(x)$ and $R(x)$ in $F[x]$ such that

$$f(x) = Q(x)g(x) + R(x) \qquad \textbf{(3-4)}$$

where $\deg R(x) < \deg g(x)$.

Rather than to give a proof of this theorem, which involves nothing more than a wordy description of the process of long division, we shall look at some examples.

$$\begin{aligned} f(x) &= x^3 + 4 \\ g(x) &= x^6 + 3x + 9 \\ F &= \text{the rational field} \end{aligned} \tag{3-5}$$

Here the quotient $Q(x)$ and remainder $R(x)$ can be immediately taken as 0 and $f(x)$ respectively, i.e.,

$$x^3 + 4 = 0(x^6 + 3x + 9) + (x^3 + 4)$$

is already of the form of Equation (3-4).

$$\begin{aligned} f(x) &= x^3 + 4 \\ g(x) &= 3x^2 + x + 2 \\ F &= \text{the rational field} \end{aligned} \tag{3-6}$$

In this example long division is required to find the quotient $Q(x)$ and the remainder $R(x)$. Such a calculation shows

$$x^3 + 4 = \underbrace{\left[\frac{1}{3}x - \frac{1}{9}\right]}_{Q(x)}(3x^2 + x + 2) + \underbrace{\left[-\frac{5}{9}x + \frac{38}{9}\right]}_{R(x)}$$

$$\begin{aligned} f(x) &= x^3 + [4] \\ g(x) &= [3]x^2 + x + [2] \\ F &= (I/5) \end{aligned} \tag{3-7}$$

Again, long division is necessary here, but the calculations are done in the field $I/(5)$. We find

$$\begin{array}{r|l} & [2]x + [1] \\ \hline [3]x^2 + x + [2] & x^3 \qquad\qquad\qquad + [4] \\ & x^3 + [2]x^2 + [4]x \\ \hline & -[2]x^2 - [4]x + [4] \\ & [3]x^2 + \quad x + [2] \\ \hline & [2] \end{array}$$

or

$$x^3 + [4] = ([2]x + [1])([3]x^2 + x + [2]) + [2]$$

Remark 3-1. *Remark on notation.* When dealing with polynomials in $F[x]$ where $F = I/(p)$ and p is a prime, it is awkward to continue denoting the coefficients of these polynomials by square brackets as in Equation

(3-7). Therefore, we shall adopt the practice of simply omitting the brackets when writing down elements from the field $I/(p)$ whenever it is clear in the context that $I/(p)$ is the field in which the computations are made. Thus, in a statement such as "Consider the polynomial $f(x) = 3x^2 + x + 2$ of $F[x]$, where $F = I/(5)$," it will be understood that the coefficients of $f(x)$ are really equivalence classes, i.e., that $f(x) = [3]x^2 + [1]x + [2]$. The reader should not right away attribute any strange looking arithmetic to a typographical error. For example, for the field $I/(5)$, $(2x + 1)(3x + 1) = x^2 + 1$ because $(2x + 1)(3x + 1) = 6x^2 + 5x + 1$ and $6 = 1$ and $5 = 0$ in $I/(5)$.

3-2 Roots of Polynomials

If for a field F and a polynomial $f(x) = a_0 + a_1x + \ldots + a_nx^n$ of $F[x]$ we substitute for x a quantity β of F, then we obtain an element $f(\beta) = a_0 + a_1\beta + \ldots + a_n\beta^n$ of F. More generally, if the field F is contained in a larger field K and β belongs to K, then $f(\beta)$ is an element of K. A *root* of $f(x)$ is a quantity β, belonging either to F or a larger field K, for which $f(\beta) = 0$.

The study of roots of polynomials is quite important in algebra — indeed, important ideas in the theories of groups and fields arose from efforts to find formulas giving the roots of polynomials of any degree.* In a later chapter on the theory of fields, we will do a rather intensive study of the roots of polynomials. This section deals with some of the more basic facts about roots.

The starting point for the study of roots is the following:

Theorem 3-2. *The Factor Theorem.* If a polynomial $f(x)$ with coefficients from a field F has a root β in F, then there exists a polynomial $Q(x)$ in $F[x]$ such that

$$f(x) = (x - \beta)Q(x) \tag{3-8}$$

Proof: At any rate, dividing $f(x)$ by the degree one polynomial $x - \beta$ we have

$$f(x) = (x - \beta)Q(x) + R$$

for a quotient $Q(x)$ and remainder R, and as the degree of R is less than one, R is necessarily a constant. Putting $x = \beta$ in the preceding equation, we find this constant R is zero, whence, Equation (3-8).

---

*For an intriguing account of the turbulent lives of the pioneers of these theories see Ore's *Niels Hendrik Abel*. University of Minnesota Press, Minneapolis, 1957.

The Factor Theorem provides an algorithm for finding the roots of a polynomial $f(x)$ of degree n. For if β is a root of $f(x)$, then the polynomial $Q(x)$ of Equation (3-8) has degree $n - 1$, and the roots of $f(x)$ which are different from β will appear among the roots of $Q(x)$. It may happen that β is also a root of $Q(x)$, in which case $Q(x) = P(x)(x - \beta)$ for some $P(x)$ in $F[x]$; whence, $f(x) = (x - \beta)^2P(x)$. Generally, β is called a *multiple* root of $f(x)$ if for some polynomial $P(x)$ in $F[x], f(x) = (x - \beta)^rP(x)$, where r is an integer ≥ 2. If in addition $P(\beta) \neq 0$, then the root β is said to be of *multiplicity* r. Thus, the multiplicity of a root β is the highest power of the factor $x - \beta$ dividing $f(x)$. Sometimes we shall call β a double root (etc.) if its multiplicity is two (etc.).

The multiple roots (if any) of a polynomial $f(x)$ in $F[x]$ may be found by introducing a formal differentiation process in the ring $F[x]$. This is done without using limits by simply defining the derivative $f'(x)$ of the polynomial

$$f(x) = a_0 + a_1x + \ldots + a_nx^n$$

to be

$$f'(x) = a_1 + 2a_2x + \ldots + na_nx^{n-1}$$

As with the ordinary derivative of calculus, the operation of differentiation obeys the following properties:

$$(f + g)' = f' + g' \qquad \text{for } f \text{ and } g \text{ in } F[x] \tag{3-9}$$

$$(af)' = af' \qquad \text{for } f \text{ in } F[x] \text{ and } a \text{ in } F \tag{3-10}$$

$$(fg)' = fg' + f'g \qquad \text{for } f \text{ and } g \text{ in } F[x] \tag{3-11}$$

It is the product rule [Equation (3-11)] which gives a test for finding the multiple roots of $f(x)$. For if β is a multiple root of $f(x)$, then $f(x) = (x - \beta)^rP(x)$ for some $P(x)$ in $F[x]$ and integer $r \geq 2$. Taking the derivative,

$$f'(x) = r(x - \beta)^{r-1}P(x) + (x - \beta)^rP'(x) \tag{3-12}$$

Setting $x = \beta$ in Equation (3-12), we find $f'(\beta) = 0$. Thus, we have the *multiple root test:* If β is a multiple root of $f(x)$, then $f(\beta) = f'(\beta) = 0$.

3-3 Factorization of Polynomials

In Chapter 1 we developed a theory of divisibility for integers, and this theory was based on the division algorithm for integers. Since there is a similar algorithm for polynomials over a field, this theory carries over to polynomials as well. In this section we point out the slight modifications necessary to extend the divisibility theory to polynomials. No proofs will be given since the results are all proved by the same methods that were used in Chapter 1.

Let us consider now a field F and the polynomial ring $F[x]$. If $f(x)$ and $g(x)$ are in $F[x]$ and $f \neq 0$, then $f(x)$ is called a *divisor* of $g(x)$ if $g(x) = f(x)Q(x)$ for some $Q(x)$ in $F[x]$, and we write $f(x) \mid g(x)$ in this case. It should be noted that if a is any nonzero element of F, then $a \mid g(x)$ for all $g(x)$ in $F[x]$. Indeed, if $g(x) = b_0 + b_1x + \ldots + b_nx^n$, then $g(x) = a(a^{-1}b_0 + a^{-1}b_1x + \ldots + a^{-1}b_nx^n)$. It follows that if $f(x) \mid g(x)$, then so does $af(x) \mid g(x)$ for any nonzero element a of F. In other words, multiplication by a nonzero constant does not affect the divisibility of polynomials.

A polynomial $f(x) = a_0 + a_1x + \ldots + a_nx^n$ is said to be *monic* if its leading coefficient $a_n = 1$.

By a *greatest common divisor* of $f(x)$ and $g(x)$ in $F[x]$ we mean a monic polynomial $d(x)$ of largest degree which simultaneously divides $f(x)$ and $g(x)$. As for integers, we write $d(x) = (f(x),g(x))$ for the greatest common divisor. Analogously to integers, we have the following proposition:

Proposition 3-2. Let $d(x)$ be a greatest common divisor of polynomials $f(x)$ and $g(x)$ in $F[x]$. Then there exist polynomials $A(x)$ and $B(x)$ in $F[x]$ such that

$$d(x) = A(x)f(x) + B(x)g(x)$$

It follows from this proposition that if $d(x)$ is a greatest common divisor of $f(x)$ and $g(x)$, then $d(x)$ is divisible by any common factor of $f(x)$ and $g(x)$. In particular, if $d_1(x)$ is any other greatest common divisor of $f(x)$ and $g(x)$, then $d_1(x) \mid d(x)$, and conversely $d(x) \mid d_1(x)$. Consequently, $d(x)$ and $d_1(x)$ are of equal degree; so, $d(x) = ad_1(x)$ for some a in F. However, as $d(x)$ and $d_1(x)$ are both monic, it follows that $a = 1$ and $d(x) = d_1(x)$. That is, the greatest common divisor of $f(x)$ and $g(x)$ is a unique polynomial.

To calculate the greatest common divisor $d(x)$ of polynomials $f(x)$ and $g(x)$, or to find the polynomials $A(x)$ and $B(x)$ of Proposition 3-2, one uses the Euclidean algorithm, just as in the case of integers. However, if the last nonzero remainder in the algorithm is not a monic polynomial, then it must be multiplied by a suitable element a of F in order to get a monic polynomial (which then becomes the greatest common divisor). As an example of this procedure, consider $d(x) = (2x^3 + x^2 + 2x + 1,$

$2x^4 + x^3 - 2x - 1)$ over the rational field F. Applying the Euclidean algorithm in this case yields

$$2x^4 + x^3 - 2x - 1 = x(2x^3 + x^2 + 2x + 1) + (-2x^2 - 3x - 1)$$
$$2x^3 + x^2 + 2x + 1 = (-x + 1)(-2x^2 - 3x - 1) + (4x + 2)$$
$$-2x^2 - 3x - 1 = \left(-\frac{1}{2}x - \frac{1}{2}\right)(4x + 2)$$

The last nonzero remainder in this process is $4x + 2$, which is not monic. Therefore, $d(x) = (1/4)(4x + 2) = x + (1/2)$ is the greatest common divisor.

A polynomial $p(x)$ of degree at least one in $F]x[$ is called *irreducible* if it cannot be written as a product of two polynomials in $F[x]$ of smaller degree. Quite often we shall say $p(x)$ is *irreducible over F*. Irreducible polynomials are much like primes and the determination of the irreducible polynomials over a given field F is an important problem. (We shall study this problem for certain fields later.) One important property of an irreducible polynomial $p(x)$ is that if $p(x) \mid f(x)g(x)$, then either $p(x) \mid f(x)$ or $p(x) \mid g(x)$. Indeed, this fact may be derived from Proposition 3-2, as for integers. This property was the key to the proof of the Fundamental Theorem of Arithmetic. Consequently, for polynomials we have the following theorem:

Theorem 3-3. Each nonconstant polynomial $f(x)$ in $F[x]$ can be written in the form

$$f(x) = ap_1(x) \ldots p_k(x) \qquad \textbf{(3-13)}$$

where a belongs to the field F and the $p_i(x)$'s are monic irreducible polynomials in $F[x]$. Moreover, the factorization Equation (3-13) is unique.

We remark only that the element a is the coefficient of the highest power of x appearing in $f(x)$.

3-4 Congruence in the Polynomial Ring *F* [*x*]

In the ring $F[x]$, where F is a field, congruence is defined exactly as for integers. That is, let $f(x)$ be a polynomial, other than the zero polynomial, in $F[x]$ and define

$$a(x) \equiv b(x) \pmod{f(x)} \quad \text{if } f(x) \mid a(x) - b(x) \qquad \textbf{(3-14)}$$

Congruence defines an equivalence relation in $F[x]$, and if $[a(x)]$ denotes the equivalence class containing the polynomial $a(x)$ then the rules

$$[a(x)] + [b(x)] = [a(x) + b(x)] \qquad \textbf{(3-15)}$$

$$[a(x)][b(x)] = [a(x)b(x)] \qquad \textbf{(3-16)}$$

are well-defined rules of addition and multiplication, and with these rules of combination the equivalence classes become a commutative ring with an identity element. This ring is denoted by $F[x]/(f(x))$. The identity element of the ring is the equivalence class $[1]$, while the zero element is the equivalence class containing the zero polynomial. (Of course, also, $0 = [f(x)]$.)

If the polynomial $f(x)$ is of degree n, then the distinct elements of the ring $F[x]/(f(x))$ are the equivalence classes $[a_0 + a_1x + \ldots + a_{n-1}x^{n-1}]$. Indeed, if $[a(x)]$ is any element of $F[x]/(f(x))$, then because of the division algorithm, $a(x) = f(x)Q(x) + R(x)$ for some quotient $Q(x)$ and remainder $R(x)$ in $F[x]$; hence, $[a(x)] = [R(x)]$, but $\deg R(x) \leq n - 1$. Thus,

$$[a(x)] = [a_0 + a_1x + \ldots + a_{n-1}x^{n-1}]$$

for some $a_0, a_1, \ldots, a_{n-1}$ belonging to F. Let us also note that if $[a_0 + a_1x + \ldots + a_{n-1}x^{n-1}] = [b_0 + b_1x + \ldots + b_{n-1}x^{n-1}]$, then $a_0 = b_0$, $a_1 = b_1, \ldots, a_{n-1} = b_{n-1}$. Indeed, the difference $(a_0 - b_0) + (a_1 - b_1)x + \ldots + (a_{n-1} - b_{n-1})x^{n-1}$ must be divisible by the polynomial $f(x)$, which is of degree n. This is possible only if $0 = a_0 - b_0 = a_1 - b_1 = \ldots = a_{n-1} - b_{n-1}$.

A very important result about these rings is the following theorem:

Theorem 3-4. If $p(x)$ is an irreducible polynomial over the field F, then $F[x]/(p(x))$ is a field. Conversely, if $F[x]/(f(x))$ is a field, then $f(x)$ is irreducible over F.

Proof: Suppose $p(x)$ is irreducible over F and let $[a(x)]$ be a nonzero element of $F[x]/(p(x))$. Then $1 = (p(x), a(x))$; hence, there exist $b(x)$ and $c(x)$ in $F[x]$ such that

$$1 = b(x)p(x) + a(x)c(x)$$

Therefore, $a(x)c(x) \equiv 1 \pmod{p(x)}$, or $[a(x)][c(x)] = [1]$. This shows that $[a(x)]$ has a multiplicative inverse, which immediately implies $F[x]/(p(x))$ is a field.

The proof of the rest of the theorem is left as an exercise.

The fields $F[x]/(p(x))$ are of fundamental importance in the theory of polynomials and their roots, and a detailed study of these fields will be made in Chapter 5. It will be instructive at this time, however, to consider a concrete example, which will illustrate the general ideas involved in the study of these fields.

Example 3-1. Let F be the field $I/(2)$. The addition and multiplication tables for F are Tables 3-1 and 3-2.

Table 3-1

+	0	1
0	0	1
1	1	0

Table 3-2

×	0	1
0	0	0
1	0	1

In these tables we have followed the notation discussed in Section 3-1 for denoting the elements of $I/(2)$.

The polynomial $p(x) = x^2 + x + 1$ is irreducible over this field. For otherwise, $p(x)$ would factor over F as

$$p(x) = (ax + b)(cx + d)$$

where a, b, c, and d are in F and $a \neq 0$, $c \neq 0$. However, then $p(x)$ would have a root in F — explicitly, $p(-a^{-1}b) = 0$. But in fact, $p(x)$ has no root in F because F contains only the elements 0 and 1 and $p(0) = p(1) = 1$. Hence, $p(x)$ is irreducible as claimed, and $F[x]/(p(x))$ is a field because of Theorem 3-4.

The elements of $F[x]/(p(x))$ are the equivalence classes $[R(x)]$, where $R(x)$ has degree at most one. Because of this fact, it is possible to list all the elements of $F[x]/(p(x))$, for the only polynomials $R(x)$ of degree at most one in $F[x]$ are

$$R(x) = x,\ x + 1,\ 1, \text{ or } 0 \qquad \textbf{(3-17)}$$

Obviously, no two distinct polynomials of Equation (3-17) are congruent modulo $x^2 + x + 1$; so, the equivalence classes $[x]$, $[x + 1]$, $[1]$, and $[0]$

are distinct. Hence, the elements of $F[x]/(p(x))$ are precisely $[x]$, $[x+1]$, $[1]$, and $[0]$.

The addition or multiplication of these elements is done by using Equations (3-15) and (3-16) and reducing the polynomials modulo x^2+x+1. For example, $[x][x+1] = [x^2+x] = [-1] = [1]$. We find the addition and multiplication tables for $F[x]/(p(x))$ are Tables 3-3 and 3-4.

Table 3-3

$+$	$[0]$	$[1]$	$[x]$	$[x+1]$
$[0]$	$[0]$	$[1]$	$[x]$	$[x+1]$
$[1]$	$[1]$	$[0]$	$[x+1]$	$[x]$
$[x]$	$[x]$	$[x+1]$	$[0]$	$[1]$
$[x+1]$	$[x+1]$	$[x]$	$[1]$	$[0]$

Table 3-4

$\times$	$[0]$	$[1]$	$[x]$	$[x+1]$
$[0]$	$[0]$	$[0]$	$[0]$	$[0]$
$[1]$	$[0]$	$[1]$	$[x]$	$[x+1]$
$[x]$	$[0]$	$[x]$	$[x+1]$	$[1]$
$[x+1]$	$[0]$	$[x+1]$	$[1]$	$[x]$

Looking at the parts of Tables 3-3 and 3-4 just involving [0] and [1], we have Tables 3-5 and 3-6.

Table 3-5

+	[0]	[1]
[0]	[0]	[1]
[1]	[1]	[0]

Table 3-6

×	[0]	[1]
[0]	[0]	[0]
[1]	[0]	[1]

From the form of Tables 3-5 and 3-6 we may argue that the set $F' = \{[0], [1]\}$ is a field. Indeed, Tables 3-5 and 3-6 differ from those of F (see Tables 3-1 and 3-2) *only in notation.* Consequently, F' is isomorphic to F.

Since the fields F and F' are isomorphic, it is convenient to use the same notation for their elements. Thus, we shall write 0 and 1 instead of [0] and [1] respectively. By this device, F' becomes completely identical to F. If we further write θ for $[x]$, i.e.,

$$\theta = [x]$$

then, the elements of $F[x]/(p(x))$ can be written in the form

$$a + b\theta \quad \text{where } a,b \text{ are in } F \tag{3-18}$$

For example, $[x + 1] = [x] + [1] = \theta + 1$ (in the new notation). Rewriting the remaining elements, $F[x]/(p(x)) = \{0,1,\theta,\theta + 1\}$.

The element θ has special significance, namely

$$p(\theta) = 0 \tag{3-19}$$

Indeed, $p(\theta) = \theta^2 + \theta + 1 = [x]^2 + [x] + [1] = [x^2 + x + 1] = [p(x)] = 0$.

A resume of the more salient features of this example is the following:

The field $F[x]/(x^2 + x + 1)$ contains an isomorphic copy of F, and the polynomial $p(x) = x^2 + x + 1$, which had no root in F, does have a root θ in the larger field $F[x]/(x^2 + x + 1)$.

In the larger field $K = F[x]/(x^2 + x + 1)$ the polynomial $x^2 + x + 1$ is divisible by $x - \theta$, because of the Factor Theorem. By long division

$$\begin{array}{r l}
 & x + (\theta + 1) \\
x - \theta \,\big) & x^2 + x + 1 \\
 & x^2 - \theta x \\ \hline
 & (\theta + 1)x + 1 \\
 & (\theta + 1)x - \theta^2 - \theta \\ \hline
 & 1 + \theta^2 + \theta\,(=0)
\end{array}$$

Hence $p(x)$, as a polynomial in $K[x]$, factors

$$p(x) = (x - \theta)(x + \theta + 1)$$

3-5 Polynomials over the Rational Field

The problem of determining whether a given polynomial with rational coefficients is irreducible over the rational field inevitably leads to questions in the theory of numbers, as we shall presently show. While there is no criteria for generally determining when a polynomial is irreducible over the rational field, there are some quite useful partial results concerning the factorization of such polynomials.

The first of these concerns the roots of such a polynomial.

Theorem 3-5. (*The rational root test*). Let $f(x) = a_0 + a_1x + \ldots + a_nx^n$ be a polynomial with integral coefficients and suppose r/s is a rational root of $f(x)$, where r and s are relatively prime integers. Then $r \mid a_0$ and $s \mid a_n$.

Proof: Before giving the proof, we remark that by canceling the common factors of the numerator and denominator any rational number can be put as a ratio of relatively prime integers. Thus, in the statement of the theorem, the condition $(r,s) = 1$ is innocuous. If $f(r/s) = 0$, then

$$a_0 + a_1\left(\frac{r}{s}\right) + \ldots + a_n\left(\frac{r}{s}\right)^n = 0 \qquad \textbf{(3-20)}$$

and

$$-a_0s^n = a_1rs^{n-1} + \ldots + a_nr^n \qquad \textbf{(3-21)}$$

Thus, r is a factor of a_0s^n. However, $(r,s) = (r,s^n) = 1$; hence, $r \mid a_0$. Similarly from Equation (3-20)

$$-a_nr^n = a_0s^n + a_1rs^{n-1} + \ldots + a_{n-1}r^{n-1}s$$

which shows that $s \mid a_nr^n$; whence, $s \mid a_n$.

While Theorem 3-5 is formulated for polynomials with integral coefficients, it may be applied to a polynomial $g(x)$ with only rational coefficients. For if k is the least common multiple of the denominators of the coefficients of $g(x)$, then $k \neq 0$ and $g(x) = (1/k)f(x)$, where $f(x)$ has integral coefficients. Certainly $g(x)$ and $f(x)$ have the same roots, so any rational root of $f(x)$ which is found by the preceding test will also be a root of $g(x)$. For example, if $g(x) = 2x^3 + x^2 - 1/2$, then $2g(x) = 4x^3 + 2x^2 - 1$ has integral coefficients, and the possible rational roots of the latter polynomial are $\pm 1, \pm 1/2, \pm 1/4$. By trial, $1/2$ is the only rational root of $g(x)$.

A polynomial $g(x)$ with integral coefficients is called *primitive* if the greatest common divisor of its coefficients is 1. For example $4x^3 + 2x + 3$ is primitive, while $4x^3 + 2x + 16$ is not. The greatest common divisor δ of the coefficients of a polynomial $f(x)$ with integral coefficients is called the *content* of $f(x)$. Factoring the content δ from each of the coefficients of $f(x)$, we have $f(x) = \delta g(x)$, where $g(x)$ is primitive.

Theorem 3-6. The product of primitive polynomials is again a primitive polynomial.

Proof: Let us suppose

$$f(x) = a_0 + a_1 x + \ldots + a_n x^n$$

and

$$g(x) = b_0 + b_1 x + \ldots + b_m x^m$$

are primitive. Their product is

$$f(x)g(x) = a_0 b_0 + (a_0 b_1 + a_1 b_0)x + \ldots + a_n b_m x^{n+m} \quad \textbf{(3-22)}$$

For a given prime p, some coefficient of $f(x)$ is not a multiple of p (because $f(x)$ is primitive) — call the first such coefficient a_i. Thus, $p \nmid a_i$ but $p \mid a_k$ for all $k < i$. Similarly there is a coefficient b_j of $g(x)$ for which $p \nmid b_j$ but $p \mid b_k$ for $k < j$.

In the product [Equation (3-22)] the coefficient of x^{i+j} is the sum of the terms $a_r b_s$, where $r + s = i + j$. Such a term can be written $a_r b_s = a_{i+j-s} b_s$, $s = 0, \ldots, i + j$.

From the choice of i and j we have

$$a_{i+j-s} b_s \equiv 0 \pmod{p}$$

if either $s < j$ or $i + j - s < i$; hence,

$$a_{i+j-s} b_s \equiv 0 \pmod{p}$$

if either $s < j$ or $j < s$. Altogether,

$$a_{i+j-s}b_s \begin{cases} \equiv 0 \pmod{p} & \text{if } s \neq j \\ \not\equiv 0 \pmod{p} & \text{if } s = j \end{cases}$$

Therefore, the coefficient of x^{i+j} is $\equiv a_i b_j \pmod{p} \not\equiv 0 \pmod{p}$.

We have shown that for any prime p there is some coefficient of $f(x)g(x)$ which is not a multiple of p. It follows that $f(x)g(x)$ is primitive.

The importance of the last theorem lies in the following result:

Lemma 3-1. *Gauss's Lemma.* If a primitive polynomial $f(x)$ factors as a product

$$f(x) = g(x)h(x)$$

of polynomials $g(x)$ and $h(x)$ with rational coefficients, then $f(x)$ factors

$$f(x) = G(x)H(x)$$

as a product of polynomials $G(x)$ and $H(x)$ with integral coefficients.

Proof: For if α is the least common multiple of the denominators of the coefficients of $g(x)$ and β is the least common multiple of the denominators of the coefficients of $h(x)$, then

$$g(x) = \frac{1}{\alpha}G_1(x) \quad \text{and} \quad h(x) = \frac{1}{\beta}H_1(x)$$

where $G_1(x)$ and $H_1(x)$ have integral coefficients. Moreover, if δ is the content of $G_1(x)$ and δ' is the content of $H_1(x)$, then

$$G_1(x) = \delta G(x) \quad \text{and} \quad H_1(x) = \delta' H(x)$$

where $G(x)$ and $H(x)$ are primitive polynomials. Then

$$f(x) = \frac{\delta\delta'}{\alpha\beta}G(x)H(x)$$

or

$$\alpha\beta f(x) = \delta\delta' G(x)H(x) \tag{3-23}$$

Since $f(x)$ is primitive, the content of the left-hand side of Equation (3-23) is $\alpha\beta$, while the content of the right-hand side of Equation (3-23) is $\delta\delta'$ because $G(x)H(x)$ is primitive. Therefore, $\alpha\beta = \delta\delta'$ and $f(x) = G(x)H(x)$, as claimed.

Remark 3-2. Typically, Gauss's Lemma is used as follows: Suppose $f(x)$ is a polynomial with rational coefficients and we want to check whether $f(x)$ is irreducible over the rational field. If k is the least common multiple of the denominators of the coefficients of $f(x)$, then $f(x) = (1/k)F(x)$, where $F(x)$ has integral coefficients. Furthermore, if δ is the content of $F(x)$, then $f(x) = (\delta/k)G(x)$, where $G(x)$ is a primitive polynomial. Now to factor $f(x)$ we have to factor $G(x)$ and conversely. If $G(x)$ is to factor over the rational field then *a priori,*

$$G(x) = (a + bx + \ldots)(a' + b'x + \ldots)$$

where $a,b, \ldots, a',b', \ldots$ are rational numbers. *However, Gauss's Lemma allows us to assume* $a,b, \ldots, a',b', \ldots$ *are, in fact, integers.* This reduction saves an enormous amount of computation.

As an application of Gauss's Lemma, we derive next the following very useful criteria for determining irreducibility over the rational field.

Eisenstein's Criteria. Let $f(x) = a_0 + a_1x + \ldots + a_nx^n$ be a polynomial with integral coefficients, and assume there is a prime p such that

$$a_0 \equiv a_1 \equiv \ldots a_{n-1} \equiv 0 \pmod{p} \quad \text{and} \quad \begin{array}{l} a_n \not\equiv 0 \pmod{p}, \\ a_0 \not\equiv 0 \pmod{p^2} \end{array} \qquad \textbf{(3-24)}$$

then $f(x)$ is irreducible over the rational field.

Proof: After factoring out the content α of $f(x)$, we will have $f(x) = \alpha F(x)$ where the polynomial $F(x) = (1/\alpha)f(x)$ has integral coefficients and is primitive. The content α is not divisible by p because $a_n \not\equiv 0 \pmod{p}$. Therefore,

$$F(x) = b_0 + b_1x + \ldots + b_nx^n$$

also has the property that $b_0 \equiv \ldots \equiv b_{n-1} \equiv 0 \pmod{p}$, $b_n \not\equiv 0 \pmod{p}$, and $b_0 \not\equiv 0 \pmod{p^2}$.

Suppose $F(x)$ factors over the rational field, for instance

$$F(x) = (c_0 + c_1x + \ldots)(d_0 + d_1x + \ldots) \qquad \textbf{(3-25)}$$

where (by Gauss's Lemma) the coefficients $c_0,c_1, \ldots, d_0,d_1, \ldots$ are integers.

Multiplying out Equation (3-25) and comparing coefficients, we have

$$\begin{aligned} b_0 &= c_0 d_0 \\ b_1 &= c_0 d_1 + c_1 d_0 \\ &\cdot \\ &\cdot \\ &\cdot \end{aligned} \tag{3-26}$$

As $p \mid b_0$ and $p^2 \nmid b_0$, p divides exactly one of c_0 and d_0 — for example $p \mid c_0$. Proceeding to the next relation of Equation (3-26), we find $p \mid c_1$. Continuing in this way, we see that $c_0, c_1, \ldots$ are all divisible by p. However, Equation (3-25) would then imply that p divides each of the coefficients of $F(x)$, which is contrary to fact. Consequently, a factorization such as Equation (3-25) is impossible, and $F(x)$ is irreducible. Of course, this means $f(x)$ is irreducible as well.

3-6 Integral Domains

In order to study polynomials in several variables, we look next at algebraic systems which fall between the categories of rings and fields.

An *integral domain* D is a commutative ring with identity element having at least two elements and such that it has no divisors of zero. Recall that this last property states the following:

Property 3-1. If $ab = 0$ for a,b in D, then either $a = 0$ or $b = 0$.

Moreover, this implies the cancellation law:

Property 3-2. If a,b, and c are elements of D and $ac = bc$, then $c = 0$ or $a = b$.

Examples of integral domains are fields, the integers, and the polynomial rings $F[x]$, where F is a field. Generally, an integral domain is an algebraic structure which is weaker than a field. Later it will be explained how all integral domains may be embedded in fields.

Integral domains arise from the study of polynomials in several variables. For example, the expression

$$f(x,y) = 1 + \frac{1}{2}xy + x^3y + y^2 + x^2y^2$$

has coefficients belonging to the rational field. However, if we write $f(x,y)$ as a quadratic polynomial in y, i.e.,

$$f(x,y) = 1 + \left(\frac{1}{2}x + x^3\right)y + (1 + x^2)y^2$$

then, the coefficients of the y terms are not rational; rather, they belong to the integral domain $F[x]$, where F is the rational field.

For an integral domain D an expression

$$a_0 + a_1x + \ldots + a_nx^n$$

where the coefficients $a_0,a_1, \ldots, a_n$ belong to D is called a *polynomial* (over D), and the totality of such polynomials is denoted by $D[x]$. Equality, degree, addition, and multiplication of polynomials of $D[x]$ are defined in exactly the same way as for polynomials with coefficients from a field. Furthermore, we have the following proposition:

Proposition 3-3. If D is an integral domain, then so is $D[x]$ an integral domain.

Proof: The verification that $D[x]$ is a commutative ring with identity element is straightforward and, therefore, is omitted. To show that Property 3-1 is true, let $f(x)$ and $g(x)$ be nonzero polynomials of $D[x]$. Then,

$$f(x) = a_0 + a_1x + \ldots + a_nx^n$$

and

$$g(x) = b_0 + b_1x + \ldots + b_mx^m$$

where the coefficients belong to D and $a_n \neq 0$ and $b_m \neq 0$. The product $f(x)g(x)$ is

$$f(x)g(x) = a_0b_0 + (a_0b_1 + a_1b_0)x + \ldots + a_nb_mx^{n+m}$$

Since D is an integral domain, a_nb_m is nonzero; hence, $f(x)g(x)$ is also nonzero, as required.

Remark 3-3. The domain $D[x]$ is larger than D. For $D[x]$ contains the constant polynomials, but these are precisely the elements of D. On the other hand, from the definition of equality of polynomials, it is clear that x, which belongs to $D[x]$, cannot be in D. Thus, $D[x]$ is strictly larger than D.

From the preceding remark it follows that we may construct integral domains one after the other on the basis of Proposition 3-3. Starting with the field F, we may construct the integral domain $F[x]$. Then, from $F[x]$ we may construct the still larger integral domain of all polynomials

$$a_0 + a_1y + \ldots + a_ny^n$$

in an indeterminant y, where the coefficients $a_0, a_1, \ldots, a_n$ lie in the integral domain $F[x]$. (We are applying here Proposition 3-3 with $D = F[x]$, but calling the indeterminant "y" since the symbol x has already been used.) This integral domain is denoted as $F[x,y]$, and it consists of all polynomials in two variables x and y with coefficients in F. Next, we may form $F[x,y,z] = F[x,y][z]$, where z is a new indeterminant.

When dealing with polynomials in several indeterminants, it is convenient to denote them as $x_1, x_2, \ldots, x_n$, if there are n of them. Then by successive applications of Proposition 3-3, we have the integral domains $F[x_1]$, $F[x_1,x_2], \ldots$, $F[x_1,x_2, \ldots, x_n]$ where $F[x_1,x_2, \ldots, x_k] = D[x_k]$, and $D = F[x_1,x_2, \ldots, x_{k-1}]$. In this way, a polynomial in $x_1, x_2, \ldots, x_n$, with coefficients from F, is an element of the integral domain $F[x_1,x_2, \ldots, x_n]$.

We have already seen that the integral domain $F[x]$, where F is a field, carries a satisfactory theory of divisibility. But does this theory remain valid for $D[x]$, if D is only an integral domain? The whole divisibility theory was based on the division algorithm. However, for the ring $D = I$, the integers, we note the following:

Property 3-3. There is no division algorithm for $I[x]$.

Consider the polynomials $5x^2 + 1$ and $x^3 + 1$. The division algorithm would assert

$$x^3 + 1 = (5x^2 + 1)Q(x) + R(x) \qquad \textbf{(3-27)}$$

for some $Q(x)$ and $R(x)$ in $I[x]$, where $R(x)$ has degree less than two. Evidently, $Q(x)$ must be of degree one — $Q(x) = ax + b$ for some integers a and b. But, then comparing the cubic terms of Equation (3-27) we see that $1 = 5a$, which is impossible for an integer a.

We are left with the problem of investigating divisibility of irreducibility for the domain $D[x]$, knowing that the key division algorithm is no longer available. Our program for overcoming this difficulty with $D[x]$ will consist of three parts:

Step 1. Enlarge the integral domain D to a field F so that the domain $D[x]$ becomes part of $F[x]$.
Step 2. Apply the existing divisibility theory of $F[x]$ to the polynomials in $D[x]$.
Step 3. Translate the results obtained in Step 2 back to $D[x]$.

3-7 Quotient Fields

It will be shown in this section how an arbitrary integral domain D can be embedded in a field F. The field F will not contain D necessarily, but it will contain an isomorphic copy of D.

Let $\mathfrak{D}$ be the set of all ordered pairs (a,b), $b \neq 0$ of elements a,b from the integral domain D, and write

$$(a,b) \backsim (c,d)$$

if $ad = bc$. The relation $\backsim$ is an equivalence relation on $\mathfrak{D}$. Indeed, it is easy to see that $\backsim$ is reflexive and symmetric. To check transitivity, suppose $(a,b) \backsim (c,d)$ and $(c,d) \backsim (e,f)$. Then, $ad = bc$ and $cf = de$; moreover, b, d, and f are nonzero since the three pairs belong to $\mathfrak{D}$. Multiplying the first of these equations by f and using the second equation, $adf = bcf = bde$. Thus, $(af)d = (be)d$. Since the cancellation law is valid in D, this implies $af = be$, or $(a,b) \backsim (e,f)$, which is the transitivity property.

The equivalence relation $\backsim$ partitions $\mathfrak{D}$ into distinct equivalence classes. Let us denote by the symbol a/b the equivalence class containing the pair (a,b), $b \neq 0$.

Addition and multiplication of these equivalence classes will be defined by the rules

$$\frac{a}{b} + \frac{c}{d} = \frac{ad + bc}{bd} \tag{3-28}$$

and

$$\left(\frac{a}{b}\right)\left(\frac{c}{d}\right) = \frac{ac}{bd} \tag{3-29}$$

These rules are identical to those for adding and multiplying rational numbers, and the process of expanding D to a field F is just a paraphrase of the construction of the rational numbers. However, as we are working with an arbitrary integral domain, and not with a concrete example such as the ring of integers, it is necessary to take some care in checking that this procedure works.

Fact 3-1. The addition and multiplication rules are well-defined.*

Let us check, for example, that multiplication is well-defined. What has to be shown is that if $a/b = a'/b'$ and $c/d = c'/d'$, then, still (a'/b')

---

*See the remark following Equation (2-4).

$(c'/d') = (ac)/(bd)$. Now from the rule Equation (3-29), $(a'/b')(c'/d') = (a'c')/(b'd')$; hence, it must be shown that $(a'c')/(b'd') = (ac)/(bd)$, or $(a'c')(bd) = (b'd')(ac)$. However, as $a/b = a'/b'$ and $c/d = c'/d'$, $ab' = ba'$ and $cd' = dc'$; thus, $(a'c')(bd) = (a'b)(c'd) = (b'a)(cd') = (b'd')(ac)$, as required. The verification that the addition is well defined is left as an exercise.

Fact 3-2. The system F of all equivalence classes $a/b, b \neq 0$, is a field.

The fact that F is a commutative ring is straightforward. (Note that since D is an integral domain the closure axioms for addition and multiplication are satisfied.) The identity element of F is $1/1$ ($= a/a$ for any $a \neq 0$ in D). Also, if a/b is a nonzero element of F, then $a \neq 0$ and $b \neq 0$, so that b/a is also in F. Then, $(a/b)(b/a) = 1/1$; thus, a/b has the multiplicative inverse b/a. Altogether, then, F is a field.

The field F we have constructed from D is called the *quotient field* of D.

Fact 3-3. The quotient field F contains an isomorphic copy of D.

The isomorphic copy of D is the set D^* of those elements of F of the form $a/1$, $a \in D$. Indeed, the mapping $f: D \longrightarrow D^*$ defined by $f(a) = a/1$ for a in D is one-to-one and onto. Also, $f(a + b) = f(a) + f(b)$ and $f(ab) = f(a)f(b)$ because of Equations (3-28) and (3-29). Hence, f is an isomorphism.

It is in the sense of Fact 3-3 that we say the integral domain D can be embedded in its quotient field F. In practice there is no difference in working with D or D^*, because we will only study those properties of integral domains which are preserved under isomorphism. Therefore, we may without loss of generality assume D is already contained in its quotient field F. This allows us to simplify the notation. For example, we shall simply write 1 for the identity element of F (instead of $1/1$) and we write a for $a/1$ if a is an element of D.

The quotient fields are the smallest ones containing integral domains, as the following will show:

Property 3-4. Any field which contains the integral domain D will contain its quotient field as well.

For if D is contained in a field K, then each $b \neq 0$ of D has an inverse b^{-1} in K. So K must contain the elements ab^{-1}, $a \in D$. But these are precisely the elements of the quotient field of D.

The integral domain $F[x]$, where F is any field, will play an important role in the theory of fields. The quotient field of this integral domain is denoted by $F(x)$, and it consists of the quotients $f(x)/g(x)$, where $f(x)$ and $g(x)$ are in $F[x]$ and $g(x) \neq 0$. $F(x)$ is called the *field of rational functions in x*. In the same way, for the integral domains $F[x_1,x_2, \ldots, x_n]$ where $x_1,x_2, \ldots, x_n$ are indeterminants, their quotient fields $F(x_1,x_2, \ldots, x_n)$ consist of the quotients $f(x_1,x_2, \ldots, x_n)/g(x_1,x_2, \ldots, x_n)$, where f and g belong to $F[x_1,x_2, \ldots, x_n]$ and g is nonzero. Again, $F(x_1,x_2, \ldots, x_n)$ is called a *rational function* field.

3-8 Divisibility Reconsidered

Since it is now possible to embed an integral domain D in its quotient field, let us now consider the problem of developing a satisfactory theory of divisibility for the integral domain $D[x]$, in accordance with the program outlined at the end of Section 3-6.

We shall assume that there is already a suitable theory of divisibility for D. Show that the theory extends to $D[x]$. Of course, it is necessary to explain in what sense an integral domain possesses such a theory.

A nonzero element a of an integral domain D is said to be a *divisor* of b in D if $b = ac$ for some c in D. A *unit* of D is an element u which has a multiplicative inverse in D. That is, there exists u^{-1} in D such that $uu^{-1} = 1$. The units of D are trivial divisors in the sense they divide everything. For if u is a unit and b is in D, then $b = u(u^{-1}b)$. In the integral domain of integers, the units are 1 and -1, while for a field F, each nonzero element of F is a unit.

An element p of an integral domain D is called *irreducible* if p is not a unit; and whenever $p = ab$ for elements a and b of D, either a or b is a unit. The irreducible elements of the integers are the primes and their negatives, while the irreducible elements of $F[x]$, where F is a field, are the irreducible polynomials.

An integral domain D is said to be a *unique factorization domain* if each nonunit $a \neq 0$ of D can be factored uniquely as a product

$$a = up_1 \ldots p_r$$

where u is a unit and $p_1, \ldots, p_r$ are irreducible elements. Examples of unique factorization domains are the integers and $F[x]$, where F is a field. Moreover, the field F itself is also a unique factorization domain since each nonzero element of F is a unit.

Evidently, if D is a unique factorization domain and $a,b, \ldots$ are elements of D, then an analogue of the greatest common divisor of these elements may be found by factoring each of $a,b,\ldots$ and forming d equal to the product of all the common irreducible factors. This common factor d will have the property that it is divisible by any common factor of $a,b,\ldots$. If the only common factors of $a,b,\ldots$ are units, then these elements are said to be *relatively prime*.

Our object is to show that if D is a unique factorization domain, then, $D[x]$ is also a unique factorization domain. Our plan for doing this is to embed D in its quotient field F and apply the results from $F[x]$ to the domain $D[x]$. What is needed to accomplish this is to be able to pass from a factorization in $F[x]$ to one in $D[x]$. In Section 3-5 we used Gauss's Lemma to do exactly this for D equal to the integers and F equal to the rational field. Therefore, for an arbitrary unique factorization domain D, let us say that a polynomial $p(x)$ in $D[x]$ is *primitive* if its coefficients are relatively prime elements of D. The *content* of an arbitrary polynomial $a(x)$ in $D[x]$ is defined to be the greatest common divisor of its coefficients, and $a(x) = \gamma p(x)$, where γ is the content of $a(x)$ and $p(x)$ is primitive.

As in the case of the integers and the rational field, we have the following two results:

Proposition 3-4. If D is a unique factorization domain and $f(x)$ and $g(x)$ are primitive polynomials of $D[x]$, then their product $f(x)g(x)$ is also primitive.

Proposition 3-5. Let D be a unique factorization domain and F be its quotient field. If $f(x)$ is a primitive polynomial of $D[x]$ and if $f(x)$ factors in $F[x]$, then $f(x)$ factors in $D[x]$.

Theorem 3-7. If D is a unique factorization domain, then so is $D[x]$.

Proof: For an arbitrary polynomial $f(x)$ of $D[x]$, $f(x) = \gamma f_1(x)$, where γ is the content of $f(x)$ and $f_1(x)$ is primitive. The polynomial $f_1(x)$ belongs to $F[x]$, where F is the quotient field of D, and in $F[x]$, $f_1(x)$ factors as a product of irreducible polynomials of $F[x]$. However, on account of the last proposition we may take these factors from $D[x]$. Therefore,

$$f(x) = \gamma p_1(x), \ldots, p_r(x)$$

where $p_1(x), \ldots, p_r(x)$ are irreducible polynomials in $D[x]$. Moreover, as D is a unique factorization domain, the content γ may be factored $\gamma = \gamma_1, \ldots, \gamma_s$ as a product of irreducible elements of D. Altogether,

$$f(x) = \gamma_1, \ldots, \gamma_s p_1(x), \ldots, p_r(x) \qquad \textbf{(3-30)}$$

is a factorization of $f(x)$ as a product of irreducible elements of $D[x]$. The factorization [Equation (3-30)] is unique to within a unit of D because D and $F[x]$ are unique factorization domains.

Corollary 3-1. If F is a field, then $F[x_1,x_2, \ldots, x_n]$ is a unique factorization domain.

Proof: As mentioned earlier, F is trivially a unique factorization domain. Hence, by repeated applications of the previous theorem, $F[x_1]$, $F[x_1,x_2], \ldots,$ and $F[x_1,x_2, \ldots, x_n]$ are all unique factorization domains.

Example 3-2. Let us factor $f(x,y) = x^4y^2 - x^5y + xy^2 - x^3y - y + x^4 - x^2y + x$ into irreducible factors in the domain $F[x,y]$, where $F =$ the rational field. Since $f(x,y)$ is only quadratic in y, it is convenient to regard $f(x,y)$ as an element of $D[y]$, where $D = F[x]$. Writing f as a polynomial in y,

$$f(x,y) = (x^4 + x)y^2 - (x^5 + x^3 + x^2 + 1)y + (x^4 + x)$$

The content of $f(x,y)$ is the greatest common divisor of its coefficients $x^4 + x$, $x^5 + x^3 + x^2 + 1$, and $x^4 + x$. Thus, the content is $x^3 + 1$ and

$$f(x,y) = (x^3 + 1)(xy^2 - (x^2 + 1)y + x)$$

Now we have to check whether the primitive polynomial $xy^2 - (x^2 + 1)y + x$ factors in $D[y]$, and in view of Proposition 3-5, it is sufficient to see whether it factors over the quotient field of $D = F[x]$. This quotient field is the rational function field $F(x)$. From the quadratic formula, we see that $xy^2 - (x^2 + 1)y + x$ has the roots x and $1/x$, which lie in $F(x)$; consequently, $xy^2 - (x^2 + 1)y + x$ factors over $F(x)$ and $F[x]$, as well. To get this factorization, we use the Factor Theorem. That is, $xy^2 - (x^2 + 1)y + x$ must be divisible by $y - x$. Explicitly,

$$xy^2 - (x^2 + 1)y + x = (y - x)(xy - 1)$$

Having factored the primitive part of $f(x,y)$, we return to its content $x^3 + 1$ and factor it in the domain $D = F[x]$. This is easy: $x^3 + 1 = (x + 1)(x^2 - x + 1)$. Therefore,

$$f(x,y) = (x + 1)(x^2 - x + 1)(y - x)(xy - 1)$$

exhibits $f(x,y)$ as a product of irreducible factors in $F[x,y]$.

Exercises

Exercise 3-1

1. For $f(x)$, $g(x) \in F[x]$ find the quotient and remainder when $g(x)$ is divided into $f(x)$ in each of the following cases:
   (a) $f(x) = x^n - 1$, $g(x) = x - 1$, $F =$ any field
   (b) $f(x) = x^6 + 3x^4 + x + 1$, $g(x) = (1/2)x^3 + 3x$, $F =$ the rational field
   (c) $f(x) = x^4 + x + 1$, $g(x) = x^2 + 1$, $F = I/(2)$.
   (d) $f(x) = x^5 + x^2 + 1$, $g(x) = 3x^3 + 2x^2 - x + 1$, $F = I/(5)$

2. For a field F consider the set F' of all infinite sequences
$$\{a_0, a_1, \ldots, a_n, \ldots\}$$
of elements $a_0, a_1, \ldots, a_n, \ldots$ of F, where all but finitely many of the a_n's are zero. Thinking of the association
$$a_0 + a_1x + \ldots + a_nx^n \leftrightarrow \{a_0, a_1, \ldots, a_n, 0, 0, \ldots\}$$
determine addition and multiplication rules for the sequences of F' in such a way that F' becomes a ring which is isomorphic to the polynomial ring $F[x]$.

Exercise 3-2

1. Using the division algorithm for polynomials, find a common root of the polynomials
$$x^4 + 10x^3 + 35x^2 + 50x + 24$$
and
$$x^3 + 13x^2 + 55x + 75$$

2. Find all multiple roots of
$$x^4 - 4x^3 + 36x - 27$$
where the preceding polynomial is taken over (1) the rational field and (2) the field $I/(2)$.

3. Let $f(x) = x^4 + ax + b$ where a and b are rational numbers. What conditions must be imposed upon a and b for $f(x)$ to have a multiple root?

4. Let $f(x) = x^4 + x^3 + x + 1$ be a polynomial over the field $I/(p)$, where p is a prime. For what primes p does $f(x)$ have a multiple root?

Exercise 3-3

1. Show that the only irreducible polynomials over the field F which have a root in F are the degree one polynomials.

2. Give an example of a polynomial $f(x) \in F[x]$ which has no root in the field F and which at the same time is not irreducible over F.

3. Suppose $f(x) \in F[x]$ has degree 2 or 3. Show that $f(x)$ is irreducible over the field F if and only if $f(x)$ has no root in F.

4. Let F be the field of all real numbers of the form $a + b\sqrt{2}$ where a and b are rational numbers. For which integers k is $f(x) = x^2 + kx + 1$ irreducible over F?

5. Find the greatest common divisor $(f(x), g(x))$ for the following polynomials of $F[x]$:
   (a) $(x^3 + x^2 + 2,\ 2x^2 + 1)$, $F = I/(3)$
   (b) $(x^4 + x^3 - 2x^2 + 2x - 3,\ x^5 + 2x^3 + x^2 + 2)$, $F = I/(5)$
   (c) $(x^4 + 6x^3 + x + 6,\ x^3 + 7x^2 + 4x - 12)$, $F =$ the rational field

6. Let $f(x) \in F[x]$ and a and b be distinct roots of $f(x)$ in the field F. Show that $f(x)$ is divisible by $(x - a)(x - b)$.

Exercise 3-4

1. Construct a field with 9 elements.

2. Let K be the field of Example 3-1, i.e., $K = F[x]/(x^2 + x + 1)$, where $F = I/(2)$. Find a polynomial $f(x) \in K[x]$ which is irreducible over the field K.

3. Prove the converse of Theorem 3-4.

4. Show that if F is a field, then there are infinitely many irreducible monic polynomials in $F[x]$.

Exercise 3-5

1. Does the polynomial $f(x) = 8x^3 - 6x - 1$ have a rational root?

2. Show that if p is a prime then the *cyclotomic* polynomial
$$f(x) = 1 + x + x^2 + \ldots + x^p$$
is irreducible over the rational field. Hint: Consider the polynomial $g(x) = f(x + 1)$.

3. Suppose $f(x) = x^4 + ax + b$, where a and b are relatively prime integers. Can $f(x)$ have a multiple rational root?

4. Show that if m and n are positive integers, then $\sqrt[n]{m}$ is either an integer or an irrational number.

5. Is the polynomial $x^4 - x^2 + 1$ irreducible over the rational field?

6. Is $f(x) = x^8 + 6x^4 + 9x + 15$ irreducible over the rational field?

Exercise 3-6

1. Let D be an integral domain and R be a ring which is isomorphic to D. Show that R is also an integral domain.

2. Show that the only elements e of an integral domain with the property $e^2 = e$ are $e = 1$ and $e = 0$.

3. Show that if an integral domain D has only finitely many elements, then it is a field. Hint: If the distinct elements of D are $1, a_2, \ldots, a_n$, then what does the cancellation law say about the elements $a_i, a_2 a_i, \ldots, a_n a_i$? In this way show that a typical element a_i of D has a multiplicative inverse.

4. In the integral domain $F[x_1, \ldots, x_n]$, where F is a field, the *partial derivatives* $\delta f/\delta x_1, \ldots, (\delta f)/(\delta x_n)$ of a function $f(x_1, \ldots, x_n)$ in $F[x_1, \ldots, x_n]$ are computed just as in calculus. Show that

$$\frac{\delta^2 f}{\delta x_i \delta x_j} = \frac{\delta^2 f}{\delta x_j \delta x_i}$$

5. A polynomial $f(x_1, \ldots, x_n) \in F[x_1, \ldots, x_n]$, where F is a field, is said to be *homogeneous of degree* k if $f(ax_1, \ldots, ax_n) = a^k f(x_1, \ldots, x_n)$ for all $a \in F$, where k is an integer ≥ 0. Show that if $f(x_1, \ldots, x_n)$ is homogeneous of degree k, then

$$kf(x_1, \ldots, x_n) = \sum_{i=1}^{n} x_i \frac{\delta f}{\delta x_i}$$

6. Show that any nonconstant divisor of a homogeneous polynomial is itself homogeneous.

Exercise 3-7

1. Verify that the addition rule [Equation (3-28)] is well defined.

2. In an integral domain D a *domain of positivity* is a set P of elements of D such that

$$a,b \in P \Rightarrow ab \quad \text{and} \quad a + b \text{ are in } P$$

and for each $x \in D$, exactly one of the following three possibilities holds: (1) $x \in D$; (2) $-x \in D$; (3) $x = 0$. An *ordered* integral domain is one which has a domain of positivity. Show that if D is an ordered integral domain, then so is its quotient field ordered.

3. Show that isomorphic integral domains have isomorphic quotient fields.

4. Let R be a commutative ring and S be the set of nonzero divisors of R, i.e.,

$$S = \{r \in R \mid ra = 0 \Rightarrow a = 0\}$$

Show that R can be embedded in a ring R' in such a way that each $r \in S$ has a multiplicative inverse in the ring R'.

Exercise 3-8

1. For the rational field F write each of the following polynomials of $F[x,y]$ as a product of irreducible polynomials:

(a) $y^2 - xy - x^2y + x^3$

(b) $x^2y^2 + xy + x^3y + x^2y + x^2 + x$

2. Let F be a field and $f(x,y)$ $g(x,y) \in F[x,y]$ be relatively prime polynomials. Show that there exist polynomials $A(x,y)$ and $B(x,y)$ in $F[x,y]$ and a nonzero polynomial $C(x) \in F[x]$ such that

$$C(x) = A(x,y)\,f(x,y) + B(x,y)g(x,y)$$

Hint: Consider $f(x,y)$ and $g(x,y)$ as elements of $D[y]$, where $D = F[x]$ and pass to the quotient field $F(x)$ of D.

3. Let $f(x,y)$ and $g(x,y)$ be relatively prime polynomials of $F[x,y]$, where F is a field. Show that $f(a,b) = g(a,b) = 0$ for only finitely many $a,b \in F$.

4

The Theory of Fields

4-1 Extensions of Fields

The theory of fields is inexorably tied to questions about polynomials and their roots. The following theorem is fundamental for understanding the structure of fields:

> **Theorem 4-1.** *The Theorem of Kronecker.* Let F be a field and $p(x)$ be an irreducible polynomial over F. Then: (a) The field $K = F[x]/(p(x))$ contains an isomorphic copy of F. (b) The polynomial $p(x)$ has a root θ in K. (c) If $p(x)$ has degree n, then each element of K can be written uniquely in the form $c_1 + c_2\theta + \dots + c_n\theta^{n-1}$, where $c_1, c_2, \dots, c_n$ are in F.

Proof: We have shown earlier that K is a field. (See Theorem 3-4.) If for a polynomial $f(x) \in F[x]$ we let $[f(x)]$ denote the equivalence class of polynomials which are congruent to $f(x)$ modulo $p(x)$, then the set F^* of equivalence classes $[a]$ containing the constant polynomials $f(x) = a$ is a field because of the addition and multiplication rules for these equivalence classes. Moreover, the mapping $\psi : F \to F^*$, which is defined by

$$\psi(a) = [a]$$

satisfies

$$\psi(a + b) = \psi(a) + \psi(b)$$

and

$$\psi(ab) = \psi(a)\psi(b)$$

for all a,b in F because of the addition and multiplication rules for the equivalence classes.

Furthermore, if $\psi(a) = \psi(b)$ for a,b in F, then $[a - b] = 0$; hence, the constant polynomial $f(x) = a - b$ is divisible by $p(x)$. Clearly, this is possible only if $a = b$. Therefore, ψ is one-to-one. The mapping ψ is certainly onto; hence, K contains the field F^*, which is isomorphic to F.

As in Example 3-1, we shall drop the brackets in describing the elements of F^*. In this way the polynomial $p(x) = a_0 + a_1 x + \ldots + a_n x^n$ may be regarded as a polynomial whose coefficients $a_0, a_1, \ldots, a_n$ lie in the field K. Putting $\theta = [x]$, we have $p(\theta) = a_0 + a_1[x] + \ldots + a_n[x]^n = [a_0] + [a_1][x] + \ldots + [a_n][x^n] = [a_0 + a_1 x + \ldots + a_n x^n] = p(x) = 0$. Thus, θ is a root of $p(x)$.

Reducing a polynomial $b(x) \in F[x]$ to its remainder $r(x)$ after division by $p(x)$, we have $[b(x)] = [r(x)]$, where $\deg r(x) < \deg p(x)$, so that

$$[b(x)] = [c_1 + c_2 x + \ldots + c_n x^{n-1}]$$

for some $c_1, \ldots, c_n \in F$. Then $[b(x)] = [c_1] + [c_2][x] + \ldots + [c_n][x]^{n-1} = c_1 + c_2\theta + \ldots + c_n\theta^{n-1}$ according to our convention on notation. Hence, each element $u = [b(x)]$ of K is of the type

$$u = c_1 + c_2\theta + \ldots + c_n\theta^{n-1}$$

If in addition

$$u = d_1 + d_2\theta + \ldots + d_n\theta^{n-1}$$

for $d_1, d_2, \ldots, d_n \in F$, then $0 = (c_1 - d_1) + (c_2 - d_2)\theta + \ldots + (c_n - d_n)\theta^{n-1} = [(c_1 - d_1) + (c_2 - d_2)x + \ldots + (c_n - d_n)x^{n-1}]$. However, then the polynomial $(c_1 - d_1) + (c_2 - d_2)x + \ldots + (c_n - d_n)x^{n-1}$ must be divisible by $p(x)$, which has higher degree. This is possible only if $0 = c_1 - d_1 = c_2 - d_2 = \ldots = c_n - d_n$. Therefore, the representation condition (c) of Theorem 4-1 for the elements of K is unique.

Remark 4-1. The properties (b) and (c) of Theorem 4-1 are useful for making computations in the field K. For example, consider the field $F = I/(2)$ and $K = F[x]/(x^2 + x + 1)$. Each element u of K is uniquely expressible $u = a + b\theta$ for a,b in F, where $\theta^2 + \theta + 1 = 0$, or $\theta^2 = -\theta - 1$. To find the inverse of for instance $1 + \theta$, put $u = (1 + \theta)^{-1} = a + b\theta$. It is required that $1 = (1 + \theta)(a + b\theta) = a + (a + b)\theta + b\theta^2 = a + (a + b)\theta + b(-\theta - 1) = (a - b) + a\theta$. Evidently, $a = 0$ and $b = -1$. That is, $(1 + \theta)^{-1} = -\theta$.

A field K is said to be an *extension* of a field F if K contains an isomorphic copy of F.

The Theorem of Kronecker is important because it gives a solution to the following problem: Given a field F and a polynomial $f(x) \in F[x]$ of degree ≥ 1, find an extension K of F which contains a root of $f(x)$. Indeed, if $p(x)$ is an irreducible factor of $f(x)$, then according to the Theorem of Kronecker, $p(x)$ has a root θ in the extension field $K = F[x]/[p(x)]$. Of course, θ will be a root of $f(x)$ as well.

A little reflection shows that by repeated application of the Theorem of Kronecker, we may find an extension K of a field F such that K contains all the roots of a given polynomial $f(x) \in F[x]$.

Example 4-1. Let F = the rational field and $f(x) = x^4 - x^3 - 2x + 2$. We shall construct an extension of F which contains all the roots of $f(x)$.

By inspection, $f(1) = 0$; hence, $f(x)$ is divisible by $x - 1$ in $F[x]$. Carrying out this division, we find

$$f(x) = (x - 1)(x^3 - 2)$$

The polynomial $x^3 - 2$ has no rational root; hence, it is irreducible over F.

In the extension $K = F[x]/(x^3 - 2)$, $x^3 - 2$ has the root $\theta = [x]$. Dividing $x^3 - 2$ by $x - \theta$ in $K[x]$, we find

$$x^3 - 2 = (x - \theta)(x^2 + \theta x + \theta^2)$$

Is $x^2 + \theta x + \theta^2$ irreducible over K? If $x^2 + \theta x + \theta^2$ does factor over K, then it is the product of two polynomials of degree one with coefficients from K, and each of these two factors must have a root in K. Now any root α of $x^2 + \theta x + \theta^2$ which lies in K is expressible in the form

$$\alpha = a + b\theta + c\theta^2 \tag{4-1}$$

where a, b, and c are in F. Putting $\alpha^2 + \theta\alpha + \theta^2 = 0$ (because α is assumed to be a root), we have

$$\begin{aligned} 0 &= (a + b\theta + c\theta^2)^2 + \theta(a + b\theta + c\theta^2) + \theta^2 \\ &= (a^2 + 2ab\theta + b^2\theta^2 + c^2\theta^4 + 2ac\theta^2 + 2bc\theta^3) \\ &\quad + (a\theta + b\theta^2 + c\theta^3) + \theta^2 \end{aligned}$$

This equation can be reduced to one involving no more than the second power of θ by using the fact that $\theta^3 = 2$ and $\theta^4 = 2\theta$. Carrying out this reduction, we find

$$\begin{aligned} 0 = (a^2 + 4bc + 2c) + (2ab + 2c^2 + a)\theta \\ + (b^2 + 2ac + b + 1)\theta^2 \end{aligned} \tag{4-2}$$

Now Equation (4-2) holds only if

$$\begin{aligned} 0 &= a^2 + 4bc + 2c \\ 0 &= 2ab + 2c^2 + a \\ 0 &= b^2 + 2ac + b + 1 \end{aligned} \tag{4-3}$$

Multiplying the first equation of (4-3) by a, the second equation by $2c$, and then subtracting these equations yields $a^3 - 4c^3 = 0$. If $c \neq 0$, then $(a/c)^3 = 4$. However, using the Rational Root Test we may easily check that there is no rational number r such that $r^3 - 4 = 0$. Since a/c is rational, we conclude, therefore, that $c = 0$. However, in this case the equations (4-3) imply $a = 0$; whence, $b^2 + b + 1 = 0$. But then b would be a rational root of $x^2 + x + 1$, which is impossible.

We have shown there are no rational numbers a, b, and c satisfying (4-3). Consequently, $x^2 + \theta x + \theta^2$ has no root in the field K, and it is, therefore, irreducible over K.

The polynomial $g(x) = x^2 + \theta x + \theta^2$ does have a root $\phi = [x]$ in the extension $L = K[x]/(x^2 + \theta x + \theta^2)$. Therefore, $x - \phi$ divides $g(x)$ in $L[x]$. Explicitly, by long division, we find $x^2 + \theta x + \theta^2 = (x - \phi)(x + \theta + \phi)$.

Altogether, the original polynomial $f(x) = x^4 - x^3 - 2x + 2$ factors over L as

$$f(x) = (x - 1)(x - \theta)(x - \phi)(x + \theta + \phi) \tag{4-4}$$

The elements of L can be described in terms of θ, ϕ and rational numbers. For each element u of L is expressible in the form

$$u = s + t\phi$$

where s and t are in K. The elements s and t are each representable as

$$s = a + b\theta + c\theta^2$$
$$t = d + e\theta + m\theta^2$$

where a, b, c, d, e, and m are rational numbers. Therefore,

$$u = a + b\theta + c\theta^2 + d\phi + e\theta\phi + m\theta^2\phi \qquad \textbf{(4-5)}$$

In L computations are carried out by first representing the elements as in Equation (4-5) and using the relations

$$\theta^3 = 2 \qquad \phi^2 = -\theta^2 - \theta\phi \qquad \textbf{(4-6)}$$

If a field K contains all the roots $\theta_1, \ldots, \theta_k$ of a polynomial $f(x) = a_0 + a_1x + \ldots + a_nx^n$ of degree n, then $f(x)$ splits over K as a product of linear factors

$$f(x) = a_n(x - \theta_1), \ldots, (x - \theta_k) \qquad \textbf{(4-7)}$$

Comparing the degrees of both sides of Equation (4-7) we see that $f(x)$ has n roots, though some of these may be repetitions.

Example 4-1 shows how one actually finds a field K in which a given polynomial $f(x)$ factors as in Equation (4-7). The proof that such a field always exists is not difficult.

Theorem 4-2. If $f(x)$ is a polynomial of $F[x]$, where F is a field, then there is an extension K of F which contains all the roots of $f(x)$.

Proof: We argue by induction on the degree of f. If $f(x) = ax + b$ is of degree one, then we may choose $K = F$ since F contains the (only) root $-ba^{-1}$ of $f(x)$.

Assuming the theorem is true for all polynomials of degree $n - 1$ over arbitrary fields, let $f(x)$ have degree n. Choosing an irreducible factor $p(x)$ of $f(x)$, the field $L = F[x]/[p(x)]$ will contain a root θ of $p(x)$. Since $f(\theta) = 0$, $f(x) = (x - \theta)Q(x)$ for some $Q(x) \in L[x]$ of degree $n - 1$. There is an extension K of L containing all the roots of the degree $n - 1$ polynomial $Q(x)$. Such an extension will contain all the roots of $f(x)$.

4-2 The Complex Field

A classic illustration of the Theorem of Kronecker is the construction of the complex number field C from the real field R, for $x^2 + 1$ is irreducible over R and $C = R[x]/(x^2 + 1)$.

In fact, the complex numbers were constructed in exactly this way by Cauchy, and Kronecker's Theorem resulted from the simple observation that it was the irreducibility of $x^2 + 1$ that was essential in the construction of the extension field C.

In the field C it is usual to write $[x] = i$; hence, each element z of C is uniquely expressible $z = a + bi$, where a and b are real numbers. The corresponding rules for addition and multiplication in C are, therefore,

$$(a + bi) + (c + di) = (a + c) + (b + d)i$$
$$(a + bi)(c + di) = (ac - bd) + (ad + bc)i$$

for real numbers a, b, c, and d.

For a complex number $z = a + bi$, $a,b \in R$, its complex *conjugate* $\bar{z}$ is defined to be $\bar{z} = a - bi$. We have the following properties for the conjugate:

$$\bar{\bar{z}} = z,\ \overline{z_1 + z_2} = \bar{z}_1 + \bar{z}_2,\ \overline{z_1 z_2} = \bar{z}_1\bar{z}_2, \text{ and } z = \bar{z} \Leftrightarrow z \tag{4-8}$$

is real.

Using (4-8) we may prove the following theorem:

Theorem 4-3. If $f(x)$ is a polynomial with real coefficients and if z is a complex root of $f(x)$, then $\bar{z}$ is also a root of $f(x)$.

For if $f(x) = a_0 + a_1x + \ldots + a_nx^n$, where $a_0, a_1, \ldots, a_n$ are real, then taking the conjugate of both sides of the equation $0 = f(z)$,

$$0 = \overline{a_0 + a_1z + \ldots + a_nz^n} = \bar{a}_0 + \bar{a}_1\bar{z} + \ldots + \bar{a}_n(\bar{z})^n$$
$$= a_0 + a_1\bar{z} + \ldots + a_n(\bar{z})^n = f(\bar{z})$$

The Fundamental Theorem of Algebra states the following:

Theorem 4-4. Each $f(x)$ in $C[x]$ of degree ≥ 1 has a root in C.

We shall not give a proof of this result. For a proof based on Theorem 6-1 see Redei's *Algebra*, vol 1, page 610.* It follows from Theorem 4-4

---

*Redei, L., *Algebra*, vol. 1 Pergamon Press, London. Copyright by Akádemiai Kiado, Hungary, 1967.

that the only irreducible polynomials over C are the degree one polynomials; hence, the following:

Corollary 4-1. If $f(x) = a_0 + a_1x + \ldots + a_nx^n$ is a polynomial of degree n over C, then

$$f(x) = a_n(x - \alpha_1), \ldots, (x - \alpha_n)$$

Where $\alpha_1, \ldots, \alpha_n$ are the n roots of $f(x)$ in C.

If $z = a + bi$, $a,b \in R$, is a nonzero complex number, then $a^2 + b^2 \neq 0$, and we may write

$$z = r\left(\frac{a}{r} + \frac{b}{r}i\right)$$

where $r = \sqrt{a^2 + b^2}$. Clearly, $-1 \leq a/r \leq 1$ and $-1 \leq b/r \leq 1$; hence, there is a number θ for which $a/r = \cos\theta$ and $b/r = \sin\theta$, and

$$z = r(\cos\theta + i\sin\theta) \qquad \textbf{(4-9)}$$

The form [Equation (4-9)] of z is called its *polar* or *trigonometric* form. It is valid even if $r = 0$. DeMoivre's Theorem states the following:

Theorem 4-5. If $z = r(\cos\theta + i\sin\theta)$ and $u = r'(\cos\phi + i\sin\phi)$ are in polar form, then the polar form of zu is $zu = rr'[\cos(\theta + \phi) + i\sin(\theta + \phi)]$.

The result of Theorem 4-5 follows immediately by multiplying out zu and using the addition formulas of trigonometry.

Taking $u = z$ in Theorem 4-5 we find $z^2 = r^2(\cos 2\theta + i\sin 2\theta)$, and generally, $z^n = r^n(\cos n\theta + i\sin n\theta)$ for a positive integer n. As a consequence of this rule for computing z^n we have the following:

The n roots of $f(x) = x^n - 1$, where n is a positive integer, are $z_k = \cos(2k\pi/n) + i\sin(2k\pi/n)$, $k = 0,1,\ldots,n-1$.

To see this note that $(z_k)^n = 1$ because of DeMoivre's Theorem. Hence, each $z_k, k = 0,1,\ldots,n-1$ is a root of $f(x)$. Moreover, it is easy to see that $z_0,z_1,\ldots,z_{n-1}$ are all distinct. Therefore, these n complex numbers are exactly the roots of $f(x)$. The complex numbers z_k are called the *nth roots of unity*. The root

$$\zeta = \cos\left(\frac{2\pi}{n}\right) + i\sin\left(\frac{2\pi}{n}\right)$$

generates all the nth roots of unity in the sense that $z_k = \zeta^k$ for $k = 0,1, \ldots, n-1$. Generally, a *primitive* nth root of unity is a complex number u such that each nth root of unity is a power of u. Thus, ζ is a primitive nth root of unity.

4-3 The Characteristic and Prime Fields

For a ring R and a positive integer n, we define for an element a of R a new element na by

$$na = \underbrace{a + \ldots + a}_{n \text{ terms}} \tag{4-10}$$

It is easy to see that

$$\begin{aligned} (n + m)a &= na + ma \\ n(a + b) &= na + nb \\ n(ma) &= (nm)a \end{aligned} \tag{4-11}$$

for all a,b in R and all positive integers n and m.

The definition [Equation (4-10)] can be extended to all integers n by defining $na = -(-n)a$ for n negative and $0a = 0$. The properties of (4-11) are then valid for all integers m and n.

The field $I/(p)$, where p is a prime, has the property $pa = 0$ for all $a \in I/(p)$, while for the rational field, if a is a nonzero rational number and n is a nonzero integer, then $na \neq 0$. These two examples turn out to be typical for integral domains in general. For if D is an integral domain and if there exists a nonzero integer n and a nonzero element d in D such that $na = 0$, then $0 = (n \cdot 1)a$, where 1 denotes the identity element of a; hence, $n \cdot 1 = 0$. Without loss of generality we may assume $n > 0$. Now let s be the smallest positive integer for which $s \cdot 1 = 0$. Then $s > 1$ (because D has at least two elements). If s factors, $s = tu$ for integers $t > 1$, $u > 1$, then $0 = s \cdot 1 = (tu)1 = (t \cdot 1)(u \cdot 1)$. As D is an integral domain, either $t \cdot 1 = 0$ or $u \cdot 1 = 0$. In either case we have contradicted the choice of the integer s. It follows that $s = p$, a prime. Moreover, for every $a \in D$, $pa = 0$ because $pa = (p \cdot 1)a = 0a = 0$. The integral domain D is said to have *characteristic* p in this case, and the prime p is called the characteristic of D. In the event that $na = 0$ implies either the integer $n = 0$ or the element $a = 0$, then the integral domain D is said to have *characteristic* 0.

We have proved the following theorem:

Theorem 4-6. Every integral domain has a characteristic which is either zero or a prime.

Corollary 4-2. Every field has a characteristic which is either zero or a prime.

We look next at the smallest field which a given field F can contain. Generally, a subset F' of a field F is called a *subfield* if F' is itself a field under the given addition and multiplication of the field F. For example, the complex field C contains the subfield R of all real numbers, together with the field Q of all rational numbers. We write $F' \subset F$ to indicate that F' is a subfield of F. Thus, for the complex field C, we have $Q \subset R \subset C$.

If F' is a subset of a field F such that F' contains at least two elements and

$$\begin{aligned} a,b \in F' &\Rightarrow a + b \quad \text{and} \quad ab \text{ are in } F' \\ a \in F' &\Rightarrow -a \in F' \\ b \in F' \quad \text{and} \quad b \neq 0 &\Rightarrow b^{-1} \in F' \end{aligned} \tag{4-12}$$

then, F' is a subfield of F. For if a is a nonzero element of F', then F' contains a^{-1}; hence, $1 = aa^{-1}$ belongs to F'. Thus, F' contains the identity element. The conditions of Equation (4-12) guarantee that the closure axioms and the axioms on the existence of additive and multiplicative inverses are valid for F'. The remaining axioms — associative, commutative, and distributive laws — are valid for F' because they hold for the entire field F, and F' is part of F. Thus, F' is a subfield of F.

Now if F is a field, then any subfield of F must contain the identity 1 of F; hence, it must contain $n \cdot 1$ for any integer n. It is easy to see that the set of integral multiples of 1

$$D = \{n \cdot 1 \mid n \text{ is an integer}\}$$

is an integral domain. From Theorem 4-6, it follows that either D has characteristic p, where p is a prime, or characteristic zero.

In the case that D has characteristic p, it is easy to see that

$$f : [n] \rightarrow n \cdot 1$$

is an isomorphism $f : I/(p) \rightarrow D$. Hence, in this case, D is isomorphic to $I/(p)$.

If D has characteristic zero, then the mapping

$$f : n \rightarrow n \cdot 1$$

is an isomorphism $f : I \rightarrow D$, where I is the ring of integers. Hence, D is isomorphic to I.

Thus, any subfield of F is either an extension of $I/(p)$ for some prime p or it contains an isomorphic copy of the integers. In the latter case, the subfield must contain an isomorphic copy of the quotient field of I. (See Property 3-4.)

Altogether, then, any subfield of F is either an extension of $I/(p)$ for a prime p or it is an extension of the rational field Q. Since each of these is a field, we have shown the following:

> **Theorem 4-7.** If F is a field, then the smallest subfield P contained in F is either isomorphic to $I/(p)$ for a prime p or to the rational field Q.

This smallest subfield P is called the *prime* field of F.

4-4 Simple Extensions

Let K be a field and F be a subfield of K. We want to determine the smallest subfield of K which contains F and a given element β of K. Let us denote this subfield by $F(\beta)$.

As $F(\beta)$ contains F and β we note from Equation (4-12) the following:

Property 4-1. $F(\beta)$ contains $f(\beta)$ for each $f(x) \in F[x]$.

To describe fully the subfield $F(\beta)$, we consider two mutually exclusive possibilities for β:

(i) β is a root of some nonzero polynomial $f(x) \in F[x]$.
(ii) For each nonzero polynomial $f(x) \in F[x]$, $f(\beta) \neq 0$.

The element β is said to be *algebraic* over F in the first of these cases, while in the second case, β is called *transcendental* over F.

Let us consider now the first of these possibilities, i.e., β is algebraic with respect to F. Then among all the polynomials of $F[x]$ which have β as a root, choose one of smallest degree. If we call this polynomial $p(x)$, then $p(\beta) = 0$, while $g(\beta) \neq 0$ for any nonzero polynomial $g(x) \in F[x]$ of degree smaller than that of $p(x)$. Clearly, then, the following statements are true:

Property 4-2. The polynomial $p(x)$ is irreducible over F.

Property 4-3. If $h(\beta) = 0$ for $h(x) \in F[x]$, then $h(x)$ is divisible by $p(x)$.

For dividing $p(x)$ into $h(x)$, $h(x) = p(x)Q(x) + R(x)$ for some quotient $Q(x) \in F[x]$ and remainder $R(x) \in F[x]$. Substituting $x = \beta$, we find $R(\beta) = 0$, and by the choice of $p(x)$ and the fact that deg $R(x) <$ deg $p(x)$, it follows that $R(x) = 0$.

As $p(x)$ is irreducible over F, $F[x]/(p(x))$ is a field. We now define a mapping ψ of $F[x]/(p(x))$ into $F(\beta)$ by the rule

$$\psi([f(x)]) = f(\beta) \tag{4-13}$$

where $[f(x)]$ is the equivalence class containing $f(x) \in F[x]$.

The mapping ψ is well defined, i.e., it is independent of the choice of the representative $f(x)$ of the equivalence class $[f(x)]$. For if $[f(x)] = [g(x)]$ for $f(x)$, $g(x) \in F[x]$, then $f(x) - g(x)$ is divisible by $p(x)$. Since $p(\beta) = 0$, $f(\beta) - g(\beta) = 0$, i.e., $\psi([f(x)]) = \psi([g(x)])$.

Obviously, we have

$$\psi([f(x)] + [g(x)]) = \psi([f(x)]) + \psi([g(x)]) \tag{4-14}$$

and

$$\psi([f(x)][g(x)]) = \psi([f(x)])\psi([g(x)]) \tag{4-15}$$

for every $f(x)$, $g(x) \in F[x]$.

Finally,

$$\psi([f(x)]) = \psi([g(x)]) \Rightarrow [f(x)] = [g(x)] \tag{4-16}$$

for $f(x)$, $g(x) \in F[x]$.

Indeed, if ψ maps $[f(x)]$ and $[g(x)]$ onto the same element of $F(\beta)$; then, $f(\beta) = g(\beta)$, or $f(\beta) - g(\beta) = 0$. Then Property 4-3 shows that $f(x) - g(x)$ is divisible by $p(x)$. Therefore, $[f(x)] = [g(x)]$.

Now, the mapping ψ maps the field $F[x]/(p(x))$ onto $F(\beta)$. Moreover, it follows from Equations (4-13) to (4-16) that ψ is an isomorphism. Therefore, $F(\beta)$ is isomorphic to $F[x]/(p(x))$.

We have proved the following:

Theorem 4-8. If $F \subset K$ are fields and $\beta \in K$ is algebraic over F, then $F(\beta)$ is isomorphic to $F[x]/(p(x))$ where $p(x) \in F[x]$ is a nonzero polynomial of minimum degree having β as a root.

Let us turn now to the case where β is transcendental over F, and let

$$D = \{f(\beta) \mid f(x) \in F[x]\}$$

We define a mapping $\psi : F[x] \to D$ by the rule

$$\psi : f(x) \to f(\beta)$$

Certainly ψ maps $F[x]$ onto D and

$$\begin{aligned} \psi(f(x) + g(x)) &= \psi(f(x)) + \psi(g(x)) \\ \psi(f(x)g(x)) &= \psi(f(x))\psi(g(x)) \end{aligned} \qquad \textbf{(4-17)}$$

for all $f(x), g(x) \in F[x]$. Also, ψ is one-to-one. For if $\psi(f(x)) = \psi(g(x))$, then $f(\beta) = g(\beta)$, and β is a root of the polynomial $f(x) - g(x)$. However, β is transcendental over F. Therefore, $f(x) - g(x)$ is zero, as required.

What we have shown is that the field $F(\beta)$ contains D, which is isomorphic to the integral domain $F[x]$. Therefore, D is itself an integral domain and $F(\beta)$ must contain its quotient field, which is isomorphic to $F(x)$, the rational function field. As $F(\beta)$ was the smallest subfield of K containing F and β, it follows that $F(\beta)$ is isomorphic to $F(x)$. That is, we have the following:

Theorem 4-9. If $F \subset K$ are fields and $\beta \in K$ is transcendental over F, then $F(\beta)$ is isomorphic to the rational function field $F(x)$.

Generally, if $F \subset K$ are fields and Ω is a subset of elements of K, then the smallest subfield of K containing F and each of the elements of Ω is denoted by $F(\Omega)$; and we say that $F(\Omega)$ is the field obtained from F by *adjoining* the elements of Ω to F. In the case Ω consists of a single element β, then $F(\beta)$ is called a *simple* extension of the field F. Theorems 4-8 and 4-9 show that there are exactly two types of simple extensions of a field F.

If Q is the rational field and R is the field of real numbers, then $Q(\sqrt{2})$ is an example of an algebraic extension (i.e., of the type described in Theorem 4-8). On the other hand, π is transcendental over Q; hence, $Q(\pi)$ is a transcendental extension of Q.

In calculus, arctan x is called a *transcendental* function. Why is this? If $F(x)$ is the rational function field, where F is the field of real numbers, then the connection between $F(x)$ and arctan x is that

$$\arctan x = \int \frac{1}{1 + x^2}\, dx$$

That is, arctan x is the integral of the rational function $1/(1 + x^2)$. Now the rational functions are differentiable at all but a finite number of points (namely, at all points except those for which the denominators of the rational functions vanish). Moreover, the derivative of a rational function is again a rational function.

Now if K is the set of all functions of x which are differentiable any number of times at all but a finite number of points, then from the differentiation rules for sums, products, and quotients, it is easy to show that K is a field containing $F(x)$. Examples of functions in K are arctan x, $\sin x$, e^x, and $\sqrt{x^3 + 2}$.

We have $F(x) \subset K$. Let us now adjoin an "integral" to $F(x)$. That is, let $\phi(x) \in K$ be a function such that $\phi(x) \in F(x)$ but $\phi'(x) \notin F(x)$.

We determine the nature of the simple extension $F(x)(\phi)$. There are two possibilities: ϕ is algebraic over $F(x)$ or ϕ is transcendental over $F(x)$. In the case ϕ is algebraic over $F(x)$, there will exist a monic polynomial $f(t) = t^n + a_1t^{n-1} + \ldots + a_n$ in t whose coefficients $a_1, \ldots, a_n$ lie in the field $F(x)$ such that $f(\phi) = 0$. Moreover, we may take $f(t)$ to be a polynomial of minimal degree over $F(x)$ which has ϕ as a root. If we differentiate the equation

$$\phi^n + a_1\phi^{n-1} + \ldots + a_n = 0$$

with respect to x, we obtain

$$(n\phi')\phi^{n-1} + a_1'\phi^{n-1} + \ldots + a_n' = 0$$

Thus, $g(\phi) = 0$ where $g(t) = (n\phi' + a_1')t^{n-1} + \ldots + a_n'$. Since we have assumed $\phi' \in F[x]$, the coefficients of $g(t)$ lie in $F(t)$. However, as $g(t)$ has smaller degree than $f(t)$, it follows that $g(t)$ is zero. In particular the coefficient $n\phi' + a_1' = 0$; whence, $\phi' = -a_1'/n$. But then integrating, we find $\phi = -a_1/n +$ a constant; hence, ϕ belongs to $F(x)$, contrary to assumption. Therefore, ϕ must be transcendental over $F(x)$.

We see, therefore, that the function $\phi(x) =$ arctan x is transcendental because it is not rational, but its derivative is.

4-5 The Degree of an Extension

Let $F \subset K$ be fields. Elements $u_1, \ldots, u_n$ of K are said to be *linearly dependent* over F if there exist $\alpha_1, \ldots, \alpha_n$ in F, not all equal to zero, such that

$$\alpha_1u_1 + \ldots + \alpha_nu_n = 0$$

An expression of the form $\alpha_1u_1 + \ldots + \alpha_nu_n$, where $\alpha_1, \ldots, \alpha_n$ belong to F, is called a *linear combination* of $u_1, \ldots, u_n$. Of course, for any $u_1, \ldots, u_n$ in K the linear combination $0u_1 + \ldots + 0u_n$ will be zero. However, when $u_1, \ldots, u_n$ are linearly dependent there will be a non-trivial linear combination of these elements which is zero.

Proposition 4-1. The elements $u_1, \ldots, u_n$ of K are linearly dependent over F if and only if one of $u_1, \ldots, u_n$ is expressible as a linear combination over F of the remaining elements.

Proof: For if $u_1, \ldots, u_n$ are linearly dependent, then $0 = \alpha_1 u_1 + \ldots + \alpha_n u_n$ for some $\alpha_1, \ldots, \alpha_n$ in F, where some $\alpha_i \neq 0$. For convenience, suppose $\alpha_1 \neq 0$. Then multiplying the dependence relation by α_1^{-1}, we have $0 = u_1 + \ldots + (\alpha_1^{-1}\alpha_n)u_n$, or $u_1 = (-\alpha_1^{-1}\alpha_2)u_2 + \ldots + (-\alpha_1^{-1}\alpha_n)u_n$. Thus, u_1 is a linear combination of $u_2, \ldots, u_n$.

Conversely, if for instance u_1 is a linear combination of $u_2, \ldots, u_n$, $u_1 = \beta_1 u_2 + \ldots + \beta_{n-1}u_n$ for $\beta_1, \ldots, \beta_{n-1}$ in F, then $0 = (-1)u_1 + \beta_1 u_2 + \ldots + \beta_{n-1}u_n$, and $u_1, u_2, \ldots, u_n$ are, therefore, linearly dependent.

We also have the following proposition:

Proposition 4-2. (1) If one of $u_1, \ldots, u_n$ is 0 then $u_1, \ldots, u_n$ are linearly dependent. (2) If $u_1, \ldots, u_n$ are linearly dependent; then, any larger set $u_1, \ldots, u_n, u_{n+1}, \ldots, u_m$ is also linearly dependent.

Proof: (1) Suppose $u_1 = 0$. Then $0 = 1u_1 + 0u_2 + \ldots + 0u_n$; hence, $u_1, \ldots, u_n$ are linearly dependent. (2) If $u_1, \ldots, u_n$ are linearly dependent, then $\alpha_1 u_1 + \ldots + \alpha_n u_n = 0$ for some $\alpha_1, \ldots, \alpha_n$ in F, not all $= 0$. Then $0 = \alpha_1 u_1 + \ldots + \alpha_n u_n + 0u_{n+1} + \ldots + 0u_m$, so that $u_1, \ldots, u_n, u_{n+1}, \ldots, u_m$ are linearly dependent.

Elements $u_1, \ldots, u_n$ of K are said to be *linearly independent* over F if they are not linearly dependent. We state now an easy, but useful fact:

Proposition 4-3. The following statements are equivalent: (1) $u_1, \ldots, u_n$ are linearly independent over F. (2) $\alpha_1 u_1 + \ldots + \alpha_n u_n = 0$ for $\alpha_1, \ldots, \alpha_n$ in $F \Rightarrow \alpha_1 = \ldots = \alpha_n = 0$.

If $F \subset K$ are fields, then elements $u_1, \ldots, u_n$ of K are said to *span* K if each element v of K is expressible as a linear combination $v = \alpha_1 u_1 + \ldots + \alpha_n u_n$, where $\alpha_1, \ldots, \alpha_n \epsilon F$. A *basis* for K over F is a set of elements $u_1, \ldots, u_n$ of K which span K and are linearly independent over F.

An important problem in the study of fields is: Given fields $F \subset K$, find a basis for K over F. It is not possible to solve this problem for all fields $F \subset K$. However, there are situations in which a solution is possible. The following results will reveal some features of those fields which do possess a basis. The key for these is the following theorem, which is known as Steinitz's Exchange Theorem:

Theorem 4-10. If $F \subset K$ are fields and K is spanned by s elements over F, then K can contain no more than s linearly independent elements over F.

Proof: Suppose K is spanned by $u_1, \ldots, u_s$, and let $v_1, \ldots, v_t$ be linearly independent elements of K. Now the element v_1 is a linear combination of the spanning elements $u_1, \ldots, u_s$. Thus,

$$v_1 = \alpha_1 u_1 + \ldots + \alpha_s u_s$$

for some $\alpha_1, \ldots, \alpha_s \in F$. At least one of the α_i's is nonzero. Otherwise, $v_1 = 0$, which contradicts the linear independence of $v_1, \ldots, v_t$. (See Proposition 4-2.) To simplify notation, let us assume $\alpha_1 \neq 0$. Then u_1 can be expressed as a linear combination of $v_1, u_2, \ldots, u_n$:

$$u_1 = \alpha_1^{-1} v_1 + (-\alpha_1^{-1}\alpha_2)u_2 + \ldots + (-\alpha_1^{-1}\alpha_s)u_s$$

Any element of K is a linear combination of $u_1, \ldots, u_s$, and as u_1 is a linear combination of $v_1, u_2, \ldots, u_s$, it follows that each element of K may be represented as a linear combination of the elements $v_1, u_2, \ldots, u_s$. That is, $v_1, u_2, \ldots u_s$ span K. So the original spanning set $\{u_1, \ldots, u_n\}$ has been replaced with $\{v_1, u_2, \ldots, u_s\}$, which also spans K. (This is why the theorem is called the Exchange Theorem.)

The element v_2 is expressible

$$v_2 = \beta_1 v_1 + \beta_2 u_2 + \ldots + \beta_s u_s$$

for some $\beta_1, \beta_2, \ldots, \beta_s$ in F, and it is not possible for $\beta_2 = \ldots = \beta_s = 0$, because v_1 and v_2 are linearly independent. Thus, some $\beta_i, i \geq 2$, is nonzero, and assuming in particular that β_2 is nonzero, we may write

$$u_2 = (-\beta_2^{-1}\beta_1)v_1 + \beta_2^{-1}v_2 + \ldots + (-\beta_2^{-1}\beta_s)u_s$$

as a linear combination of $v_1, v_2, u_3, \ldots, u_s$. As before, $v_1, v_2, u_3, \ldots, u_s$ span K.

Continuing in this way, we may replace the u_i's with the v_j's, and we will always obtain a set which spans K. If it were true that $t > s$, then we would exhaust the elements $u_1, \ldots, u_s$ in this process. That is, at some stage, $v_1, \ldots, v_k$, $k < t$, would span K. However, then v_{k+1} would be a linear combination of the spanning elements, contrary to the linear independence of $v_1, \ldots, v_{k+1}$. Therefore, $t \leq s$, as claimed.

Corollary 4-3. Suppose $F \subset K$ are fields and K has a basis consisting of n elements with respect to F. Then any set of $n + 1$ elements of K will be linearly dependent over F.

It is instructive to return to the Theorem of Kronecker with the information from this corollary. According to the Theorem of Kronecker, if $p(x)$ is an irreducible polynomial over the field F, then each element in the extension $K = F[x]/(p(x))$ is uniquely expressible in the form $c_1 + c_2\theta + \ldots + c_n\theta^{n-1}$, where n is the degree of $p(x)$ and $c_1, c_2, \ldots, c_n$ belong to F. Therefore, the elements $1, \theta, \ldots, \theta^{n-1}$ span the field K with respect to F. Furthermore, as each element u of K is *uniquely* expressible as

$$u = c_1 + c_2\theta + \ldots + c_n\theta^{n-1}$$

the elements $1, \theta, \ldots, \theta^{n-1}$ are linearly independent over F. For 0 is uniquely expressible as $0 = 0 \cdot 1 + 0 \cdot \theta + \ldots + 0 \cdot \theta^{n-1}$; so, this is the only linear combination of $1, \theta, \ldots, \theta^{n-1}$ which can be zero, i.e., $1, \theta, \ldots, \theta^{n-1}$ are linearly independent. Therefore, $1, \theta, \ldots, \theta^{n-1}$ form a basis for K with respect to F. If β is any element of K, then the $n + 1$ elements $1, \beta, \beta^2, \ldots, \beta^n$ must be linearly dependent over F; hence, for some $a_0, a_1, \ldots, a_n \in F$, not all $= 0$, we have $= a_0 + a_1\beta + \ldots + a_n\beta^n$. Setting $f(x) = a_0 + a_1x + \ldots + a_nx^n$, we note the following:

Property 4-4. If $p(x)$ is an irreducible polynomial of degree n over the field F, then each element β of the extension $K = F[x]/(p(x))$ is the root of some nonzero polynomial $f(x) \in F[x]$ of degree $\leq n$.

An immediate consequence of Corollary 4-3 is: If $F \subset K$ are fields such that K has a basis of n elements with respect to F, then any other basis for K will have exactly n elements.

An extension K of a field F is said to be a *finite* extension if K has a basis with respect to F. The number n of elements in such a basis is called the *degree* of the extension; and we write $n = [K : F]$ for the degree. The degree does not vary from one basis to another — it depends only on the fields K and F.

By the argument immediately preceding Property 4-4, we may show the following:

Property 4-5. If K is an extension of degree n of the field F, then any element of K is the root of a nonzero polynomial $f(x) \in F[x]$ of degree at most n.

Therefore, the following is true.

> **Theorem 4-11.** If K is a finite extension of the field F, then every element of K is algebraic over F.

If we have three fields F, K, and L, where $F \subset K \subset L$, then K is an extension of F, L is an extension of K, and L is an extension of F. What is the relation of the degrees of these various extensions? We have the following very neat answer to this question.

Theorem 4-12. *Theorem on the multiplicity of degrees.* Suppose $F \subset K \subset L$ are fields such that K is a finite extension of F and L is a finite extension of K. Then L is a finite extension of F, and $[L : F] = [L : K][K : F]$.

Proof: Let $n = [L : K]$ and $m = [K : F]$. Then we have elements $u_1, \ldots, u_n$ of L which form a basis for L with respect to K and elements $v_1, \ldots, v_m$ of K which are a basis for K over F.

The theorem will be proved by establishing that the mn elements $u_i v_j, i = 1, \ldots, n, j = 1, \ldots, m$, form a basis for L over F.

Suppose for some α_{ij}'s in F that

$$0 = \sum_{i,j} \alpha_{ij} u_i v_j$$

Then,

$$0 = (\alpha_{11} v_1 + \ldots + \alpha_{1m} v_m) u_1 + \ldots + (\alpha_{n1} v_1 + \ldots + \alpha_{nm} v_m) u_n$$

The coefficients $\alpha_{11} v_1 + \ldots + \alpha_{1m} v_m, \ldots, \alpha_{n1} v_1 + \ldots + \alpha_{nm} v_m$ in this relation belong to the field K and $u_1, \ldots, u_n$ are linearly independent over K; hence, each of these coefficients must $= 0$. That is,

$$\begin{array}{ccc} \alpha_{11} v_1 + \ldots + \alpha_{1m} v_m & = & 0 \\ \cdot & & \cdot \\ \cdot & & \cdot \\ \cdot & & \cdot \\ \alpha_{n1} v_1 + \ldots + \alpha_{nm} v_m & = & 0 \end{array} \quad \textbf{(4-18)}$$

Since $v_1, \ldots, v_m$ are linearly independent over F, in each of the equations of (4-18) we must have $\alpha_{11} = \ldots = \alpha_{1m} = 0, \ldots, \alpha_{n1} = \ldots = \alpha_{nm} = 0$. Hence, $\alpha_{ij} = 0$ for all i and j, and the elements $u_i v_j$ are linearly independent over the field F.

To complete the proof of the theorem, it is sufficient to show the elements $u_i v_j$ span L with respect to F. As $u_1, \ldots, u_n$ form a basis for L over K, any element w of K can be written

$$w = \beta_1 u_1 + \ldots + \beta_n u_n \quad \textbf{(4-19)}$$

for some $\beta_1, \ldots, \beta_n \in K$. However, using the basis $v_1, \ldots, v_m$ for K, each of the β_i's can be expressed

$$\beta_i = \alpha_{i1} v_1 + \ldots + \alpha_{im} v_m \quad \textbf{(4-20)}$$

for $\alpha_{i1}, \ldots, \alpha_{im}$ in F. Substituting the expressions of Equation (4-20) for the β_i's into Equation (4-19) and multiplying out, we see that w is actually a linear combination over F of the $u_i v_j$'s. Therefore, these elements span L with respect to F.

Using this important result on the multiplicities of extensions, the result Theorem 4-11 may be considerably sharpened.

Theorem 4-13. If L is a finite extension of a field F and if the element β of L is a root of an irreducible polynomial $f(x) \in F[x]$ of degree n, then n must divide the degree $[L : F]$ of the extension.

Proof: We have seen earlier that the simple extension $F(\beta)$ is isomorphic to $F[x]/f(x))$. (See Theorem 4-8.) According to the Theorem of Kronecker the degree of the extension $F(\beta)$ of F is $n = [F(\beta) : F]$. As $F \subset F(\beta) \subset L$, we have $[L : F] = [L : F(\beta)][F(\beta) : F]$; hence, n divides $[L : F]$.

4-6 Geometric Constructions

A famous problem in geometry is: Can an arbitrary angle be trisected using a compass and straightedge alone? Using the theory of fields, we can show that it is not possible to perform such a trisection.

The two things we can do with compass and straightedge are

(a) draw a circle of given length and given center.
(b) draw a line through two given points.

By drawing a line l, choosing a point 0 on l and marking off equidistant points along l with the compass, we may make correspond to each point of the line l an integer n (see Figure 4-1).

Using the compass and straightedge, it is possible to divide a given line segment into any number of equal parts. Doing this for each of the line segments of Figure 4-1, we may construct all points of l which have a rational distance from the origin 0 of the line l. Hence, in this sense, the field Q of all rational numbers can be constructed using compass and straightedge.

Evidently, by constructing a line l' perpendicular to l at 0 we obtain a coordinate axes (see Figure 4-2) and any point with rational coordinates in the plane may be constructed using compass and straightedge alone.

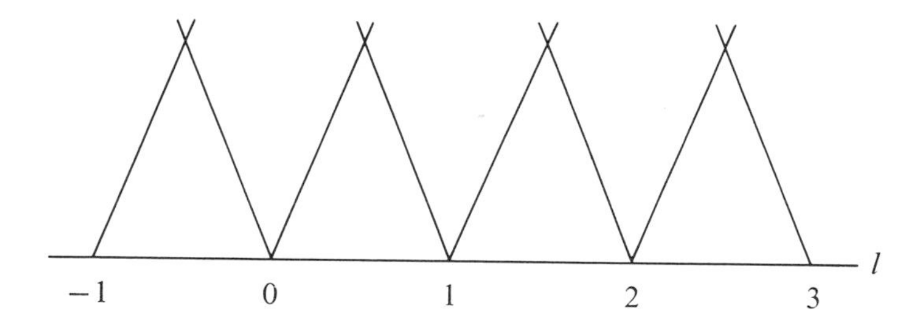

Figure 4-1

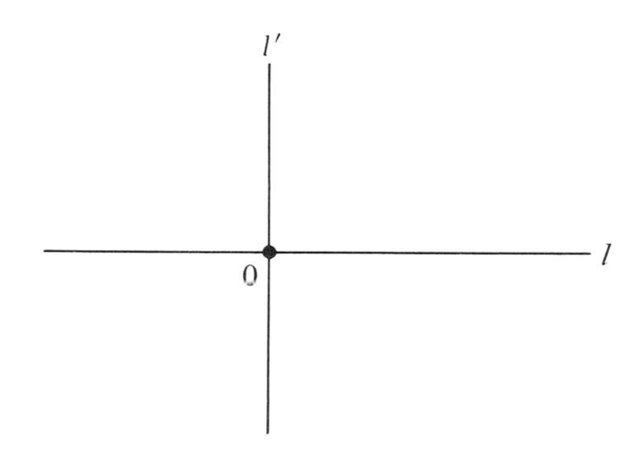

Figure 4-2

Can we construct other points? At this stage we may draw

(a) a circle of given rational radius and center with rational coordinates.
(b) a line passing through any two points with rational coordinates.

Naturally, the point of intersection of a pair of circles of type (a) or two lines (b) or a line and a circle can, therefore, be constructed. Now the intersection of two lines of the type (b) will again have rational coordinates; while for the intersection of two circles or a circle and a line, the point of intersection is found by solving a quadratic equation, and this requires finding the square root of a rational number.

Therefore, if u is the x-coordinate or y-coordinate of a point that can be constructed from the rational field by compass and straightedge, then

$u \in Q(\sqrt{d})$ for some rational number d. Repeating this process, we see that generally if v is one of the coordinates of a point which can be constructed by a sequence of operations involving the compass and straightedge, then there is a corresponding sequence of subfields of the real field R

$$Q = Q_0 \subset Q_1 \subset Q_2 \subset \ldots \subset Q_n$$

such that $Q_i = Q_{i-1}(\sqrt{d_i})$ for some $d_i \in Q_{i-1}$ and $v \in Q_n$. Since $[Q_i : Q_{i-1}] = 2$, it follows from Theorem 4-11 that $[Q_n : Q] = 2^n$.

Theorem 4-14. It is not possible to trisect an angle of 60° using compass and straightedge.

Proof: In Figure 4-3, we have taken the radial line of the 60° angle to be of length one.

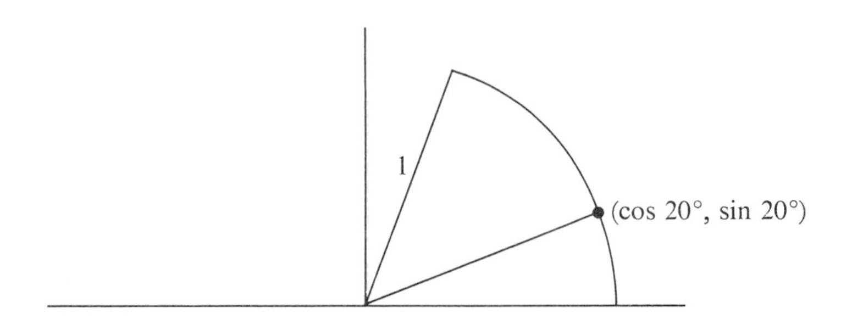

Figure 4-3

The x-coordinate of one of the trisection points is cos 20°. From trigonometry we have $\cos 3\theta = 4\cos^3\theta - 3\cos\theta$. Choosing $\theta = 20°$ and noting that $\cos 3\theta = 1/2$, we have for the coordinate $v = \cos 20°$ that $f(v) = 0$, where $f(x) = 4x^3 - 3x - 1/2$. Moreover, there is an extension Q_n of the rational field Q such that $v \in Q_n$ and $[Q_n : Q] = 2^n$ for some integer n, providing the trisection is possible.

However, the polynomial $f(x)$ is easily seen to be irreducible over Q (use the Rational Root Test); hence, by Theorem 4-12, the degree of the extension Q_n relative to Q must be divisible by 3, which it is not. Therefore, the 60° angle may not be trisected by compass and straightedge alone.

Exercises

Exercise 4-1

1. Find an extension of the rational field which contains all the roots of $x^4 - 16x^2 + 4$.

2. Find an extension of the field $I/(2)$ which contains all the roots of $x^3 + x + 1$.

3. Let $F = I/(p)$, where p is a prime, and let $f(x) \in F[x]$ be an irreducible polynomial of degree n. Determine how many elements the field $F[x]/(f(x))$ contains.

4. Show that $x^4 + 1$ is irreducible over the rational field F and factor $x^4 + 1$ as a product of irreducible polynomials over the field $K = F[x]/(x^4 + 1)$.

5. Verify that $x^2 + 1$ is irreducuble over the field $F = I/(3)$. In the affine geometry coordinatized by $K = F[x]/(x^2 + 1)$, find the point at which the lines AB and CD intersect where A,B,C, and D are the points $(1,1)$, $(\theta, 1 + \theta)$,$(2,1)$, and $(0,0)$ respectively, and $\theta = [x]$.

Exercise 4-2

1. Show there are $\phi(n)$ primitive nth roots of unity, where ϕ denotes the Euler ϕ-function.

2. Let ζ be a primitive nth root of unity and k be an integer. Show that
$$1 + \zeta^k + \zeta^{2k} + \ldots + \zeta^{(n-1)k} = n$$
if $k \equiv 0 \pmod{n}$ and
$$1 + \zeta^k + \zeta^{2k} + \ldots + \zeta^{(n-1)k} = 0$$
if $k \not\equiv 0 \pmod{n}$.

3. Show that for the real field F the only irreducible polynomials over F are: (a) polynomials of degree one and (b) polynomials $ax^2 + bx + c$ of degree two, where $b^2 - 4ac < 0$. Hint: Use Theorems 4-3 and 4-4.

4. Let z be a complex root of $f(x) = x^n - u$, where u is a complex number and n is a positive integer. Show that the n roots of $f(x)$ are $z, \zeta z, \ldots, \zeta^{n-1}z$, where ζ is a primitive nth root of unity.

5. Factor $x^6 - 1$ as a product of irreducible polynomials over (a) the complex field, (b) the real field, (c) the rational field.

6. Factor $x^4 + x^2 + 1$ as a product of irreducible polynomials over each of the fields of the previous problem.

7. Find all complex numbers z such that $z^4 + 3z^2 + z + 1 = z^3 + 2z + 2 = 0$.

Exercise 4-3

1. Show that the characteristic of an integral domain and its quotient field are the same.

2. Show that in a field of characteristic $p \neq 0$ the equation
$$(a+b)^p = a^p + b^p$$
holds for all elements a and b of the field.

3. Show that for each prime p there is a field with infinitely many elements of characteristic p.

4. Give an example of nonisomorphic fields which have isomorphic prime fields.

5. Show that isomorphic fields have isomorphic prime fields.

Exercise 4-4

1. Let $K \supset F$ be fields and $f(x) \in F[x]$ be irreducible over the field F. Show that if K contains roots α and β of $f(x)$, then $F(\alpha)$ and $F(\beta)$ are isomorphic fields.

2. Let F be the rational field. Find a real number α such that $F(\alpha) = F(\Omega)$, where $\Omega = \{\sqrt{3}, \sqrt{5}\}$.

3. Let $K \supset F$ be fields and β be an element of K which is algebraic over F. Let $f(x) \in F[x]$ be an irreducible polynomial over F which has β as a root. Show that if F has characteristic zero, then $f(x)$ has no multiple roots. Is the same thing true if F has prime characteristic?

4. Let F be a field with k elements. Show that $a^k = a$ for all $a \in F$. (Hint: Look at the proof of Euler's Theorem 2-4.)

5. Show that $\sin x$ is not a rational function of x. Then prove that it is a transcendental function of x.

6. Let F be a field and y be a nonconstant rational function of x. For the rational function field $F(x)$ prove that the element x is algebraic over the field $F(y)$.

Exercise 4-5

1. Calculate the degree of the simple extension $Q(\sqrt{2} + \sqrt{5})$ over the rational field Q.

2. Using the Fundamental Theorem of Algebra, show the only finite extensions of the real field R are R itself and the complex field C.

3. Calculate the degree of the extension $Q(\sqrt{1 + \sqrt{i}})$ of the rational field Q.

4. Let K be a field with finitely many elements. By considering the prime field P of K and a basis for K over P, show that K contains p^n elements for some prime p.

5. Let $K \supset F$ be fields and let $\beta_i \in K$ be algebraic over F for $i = 1, \ldots, n$. Prove that $F(\Omega)$ is a finite extension of F, where $\Omega = \{\beta_1, \ldots, \beta_n\}$.

Exercise 4-6

1. Show that it is not possible to construct by compass and straightedge a cube whose volume is twice that of a given cube. (If the original cube is taken to have volume $= 1$, then the equation $x^3 = 2$ gives the volume of the second cube.)

2. Given the fact that π is transcendental over the rational field, show that one cannot construct by compass and straightedge along a square whose area is that of a given circle.

5

The Theory of Groups

5-1 Definitions and Elementary Properties of a Group

Of all mathematical theories there is probably none which is so extensive in its applications as the theory of groups. The diverse subjects of geometry, combinatorial analysis, differential geometry, topology, and the theory of functions all deal at some point with the theory of groups. In addition, the theories of fields and algebraic equations are deeply based on the theory of groups. Besides deriving some of the more basic results in the theory of groups, we shall give a few applications of groups to geometry.

A *group* G is a system of objects in which any pair of objects a and b may be combined so that the following properties are true for the combined element ab:

Axiom 5-1. For all a,b in G, ab is also contained in G. (closure axiom)

Axiom 5-2. For all a,b,c in G, $(ab)c = a(bc)$. (associative law)

Axiom 5-3. G contains an element 1 such that $1a = a1 = a$ for all a in G. (existence of an identity axiom)

Axiom 5-4. For each $a \in G$ there exists a unique object b such that $ab = ba = 1$. (axiom of the existence of inverses)

The element ab is usually called the *product* of $a,b \in G$.

If the group G satisfies the following axiom, then G is called an *Abelian* group (after the Norwegian mathematician Neils Henrik Abel), or sometimes, a *commutative* group.

Axiom 5-5. $ab = ba$ for all $a,b \in G$. (commutative law)

The element 1 of Axiom 5-3 is easily seen to be unique, and it is called the *identity* of the group G. The element b of Axiom 5-4 is called the *inverse* of a and we write $b = a^{-1}$ for the inverse.

Example 5-1. The set I of all integers under the operation of addition is a group. The set F^* of all nonzero elements of a field F is a group under the operation of the multiplication of the field F. If R is any ring, then R is a group with respect to the ring addition. The set of all complex numbers z for which $z^n = 1$ is a group with respect to the multiplication of complex numbers. The set of all rotations of the Euclidean plane about the origin is a group if we define the product $\rho\rho'$ of rotations ρ and ρ' to mean the rotation resulting by first applying the rotation ρ and, then, the rotation ρ' afterwards.

Remark 5-1. In the case the group operation of a group G is denoted by $a + b$, rather than ab, it is more natural to denote the identity element of G by the symbol 0, and the inverse of $a \in G$ is written as $-a$, rather than a^{-1}.

If a group G has only finitely many elements, we call G a *finite* group. If there are n elements in G, then G is said to be of *order* n and we write $n = [G : 1]$ to denote its order.

As in the case of rings, we may form the multiplication table for a group $G = \{1,a,b, \ldots\}$ of finite order, as in Table 5-1. Table 5-1 is also

Table 5-1

	$1 \quad a \quad b \quad \cdot \quad \cdot \quad \cdot$	u	$\cdot \quad \cdot \quad \cdot$
1 a b $\cdot$ $\cdot$ $\cdot$			
v		vu	$\cdot \quad \cdot \quad \cdot$
$\cdot$ $\cdot$ $\cdot$	$\cdot \quad \cdot \quad \cdot$	$\cdot$ $\cdot$ $\cdot$	

called the *Cayley* table or *operation* table for G. A group G is completely described by its multiplication table. Tables 5-2 and 5-3 are multiplication tables of groups of order 2 and 3 respectively.

Table 5-2

	1	a
1	1	a
a	a	1

Table 5-3

	1	a	b
1	1	a	b
a	a	b	1
b	b	1	a

A group is Abelian if and only if its multiplication table is symmetric about the diagonal of the table. Thus, the groups of Tables 5-2 and 5-3 are Abelian.

For any group G we have the *cancellation laws:*

(a) $ab = ac \Rightarrow b = c$, for a,b,c in G

(b) $ba = ca \Rightarrow b = c$, for a,b,c in G

The cancellation laws (a) and (b) are called the *left* and *right* cancellation laws, respectively. In the case G is an Abelian group, (a) and (b) reduce to a single cancellation law. Let us verify law (a): If $ab = ac$, then multiplying on the left by a^{-1} and using the associative law (b), we find $(a^{-1}a)b = (a^{-1}a)c$, and as $a^{-1}a = 1$, $1b = b = 1c = c$.

Using the cancellation laws we may show the following:

In a group G the equation $ax = b$, where a and b are given elements of G, may be solved uniquely for x. Similarly, the equation $ya = b$ has a unique solution y.

The associative law shows $x = a^{-1}b$ to be a solution of $ax = b$, while if y is another solution, then $ax = ay$, and $x = y$ by the left cancellation law.

The cancellation laws are evidently reflected in the following way in the multiplication table of a group:

In no row of the multiplication table does there appear a repeated entry. Similarly, the columns of the table contain no repetitions.

This fact is useful in determining the multiplication tables of groups of small order (see Exercise 5-1, Problem 2).

Following the plan of Section 2-7 for identifying algebraic systems, we say that groups G and G' are *isomorphic* if there is a one-to-one onto mapping $f: G \to G'$ such that

$$f(ab) = f(a)f(b) \quad \text{for all } a,b \in G.$$

The mapping f is called an *isomorphism.*

Proposition 5-1. If $f: G \to G'$ is an isomorphism, then $f(1)$ is the identity of G' and $f(a)^{-1} = f(a^{-1})$

Proof: For $f(1) = f(1 \cdot 1) = f(1)f(1)$. Applying the cancellation law to $f(1) = f(1)f(1)$, it follows that $f(1)$ is the identify of G'. Then, $1 = f(1) = f(aa^{-1}) = f(a)f(a^{-1})$ shows $f(a)^{-1} = f(a^{-1})$.

When writing a product of three or more elements of a group it is immaterial where the parentheses are placed in the product to separate the factors. This is on account of the associative law. For example, $((ab)c)d = (a(bc))d = (ab)(cd) = a((bc)d) = a(b(cd))$ for a product of four elements. In writing products we shall, therefore, often omit completely such parentheses.

For an element a of a group G and a positive integer n, we define

$$a^n = \underbrace{aa \dots a}_{n \text{ factors}} \tag{5-1}$$

It is clear that

$$a^{n+m} = a^n a^m \qquad (a^n)^m = a^{nm} \tag{5-2}$$

for positive integers n and m. The exponentials a^n may be defined for all integers n by setting $a^0 = 1$ and $a^n = (a^{-n})^{-1}$ for $n < 0$. It is a simple matter to verify the law [Equation (5-2)] on exponents is true for all integers n and m.

5-2 Permutation Groups

A *permutation* of a set X is a one-to-one onto mapping $\sigma : X \to X$. For a permutation σ it is customary to write $(x)\alpha$ for the image of $x \in X$ under the mapping α, rather than the more familiar $\alpha(x)$; thus, if $X = \{1,2,3\}$ and

$$\sigma\begin{cases} 1 \longrightarrow 3 \\ 2 \longrightarrow 1 \\ 3 \longrightarrow 2 \end{cases}$$

is a permutation of X, then we write $(1)\alpha = 3$ $(2)\alpha = 1$, and $(3)\alpha = 2$ (rather than $\alpha(1) = 3$, $\alpha(2) = 1$, and $\alpha(3) = 2$).

The *product* $\alpha\beta$ of two permutations α, β of a set X, in the sense of the previous notation, is defined by

$$(x)\alpha\beta = ((x)\alpha)\beta \qquad x \in X \tag{5-3}$$

It follows that

$$(\alpha\beta)\gamma = \alpha(\beta\gamma) \tag{5-4}$$

for all permutations α,β,γ of X. Indeed, $(x)(\alpha\beta)\gamma = ((x)\alpha\beta)\gamma = (((x)\alpha)\beta)\gamma = ((x)\alpha)(\beta\gamma) = (x)(\alpha(\beta\gamma))$ according to Equation (5-4). Thus, $(\alpha\beta)\gamma$ and $\alpha(\beta\gamma)$ are identical.

Recall that the identity map 1_X of X is defined by

$$1_X : x \longrightarrow x \qquad x \in X \tag{5-5}$$

Henceforth, we shall simply write 1 for 1_X.

Proposition 5-2. The set S of all permutations of a set X is a group.

Proof: Let us show first that the product $\alpha\beta$ of permutations α and β of S is again a permutation of S. If $(x)\alpha\beta = (x')\alpha\beta$ for x,x' in X, then $((x)\alpha)\beta = ((x')\alpha)\beta$, and as β is one-to-one, $(x)\alpha = (x')\alpha$. Since α is also one-to-one, $x = x'$, i.e., $\alpha\beta$ is one-to-one. To show $\alpha\beta$ is onto, let y be any element of X. Now β maps S onto S; hence, $y = (z)\beta$ for some $z \in X$, and as α is onto, $z = (x)\alpha$ for some $x \in X$. Then $y = ((x)\alpha)\beta = (x)\alpha\beta$. That is, $\alpha\beta : x \longrightarrow y$; hence, $\alpha\beta$ is onto.

We have previously verified the associative law [Equation (5-4)] for the permutations of X, and S contains an identity element (i.e., the identity map); hence, to complete the proof that S is a group, we must show that for each $\alpha \in S$ there exists $\beta \in S$ such that $\alpha\beta = \beta\alpha = 1$. However, as α is one-to-one and onto, the existence of such a mapping β follows immediately from Theorem 2-9.

Remark 5-2. The group S of all permutations of the set X is called the *symmetric* group (with respect to the set $\overline{X}$). In the case that X contains finitely many elements, for instance n elements, then the corresponding symmetric group is denoted by S_n.

As far as the group S_n is concerned, it is immaterial how one chooses to denote the elements of the set X. Often we write $X = \{1,2, \ldots, n\}$ or even $X = \{A,B,C, \ldots\}$. A permutation α of a set $X = \{1,2, \ldots, n\}$ is completely determined once the images $(1)\alpha, (2)\alpha, \ldots, (n)\alpha$ are specified, and if we write beneath each integer k its image $(k)\alpha$, then the resulting scheme

$$\begin{pmatrix} 1 & 2 & \ldots & n \\ (1)\alpha & (2)\alpha & \ldots & (n)\alpha \end{pmatrix} \tag{5-6}$$

fully describes the permutation σ. For example, if α is the following permutation of the set $X = \{1,2,3,4,5\}$

$$\alpha \begin{cases} 1 \rightarrow 4 \\ 2 \rightarrow 3 \\ 3 \rightarrow 1 \\ 4 \rightarrow 2 \\ 5 \rightarrow 5 \end{cases}$$

then

$$\alpha = \begin{pmatrix} 1 & 2 & 3 & 4 & 5 \\ 4 & 3 & 1 & 2 & 5 \end{pmatrix}$$

In calculating the product of two permutations we are to "read" the permutations from left to right, according to Equation (5-3). Thus,

$$\begin{pmatrix} 1 & 2 & 3 & 4 & 5 \\ 4 & 3 & 1 & 2 & 5 \end{pmatrix}\begin{pmatrix} 1 & 2 & 3 & 4 & 5 \\ 4 & 2 & 5 & 3 & 1 \end{pmatrix} = \begin{pmatrix} 1 & 2 & 3 & 4 & 5 \\ 3 & 5 & 4 & 2 & 1 \end{pmatrix}$$

It is not difficult to calculate the order of the symmetric group S_n. For if α is a permutation of the set $X = \{1,2, \ldots, n\}$. then there are exactly n choices for $(1)\alpha$. This leaves $n-1$ possible values for $(2)\alpha$, and continuing in this way, we see there are $n(n-1)(n-2) \ldots 2 \cdot 1 = n!$ ways of filling in the bottom row of

$$\begin{pmatrix} 1 & 2 & & n \\ (1)\alpha & (2)\alpha & \ldots & (n)\alpha \end{pmatrix}$$

Therefore, we note the following:

Proposition 5-3. The order of the symmetric group S_n is $n!$

Let us list now the elements and Cayley table for the symmetric group S_n, $n = 1, 2, 3$. (See (5-7), (5-8), and (5-9.)

$$S_1 = \{1\} \tag{5-7}$$

$$S_2 = \left\{1 = \begin{pmatrix} 1 & 2 \\ 1 & 2 \end{pmatrix}, \alpha = \begin{pmatrix} 1 & 2 \\ 2 & 1 \end{pmatrix}\right\} \tag{5-8}$$

	1	α
1	1	α
α	α	1

$$S_3 = \left\{1, \alpha = \begin{pmatrix} 1 & 2 & 3 \\ 1 & 3 & 2 \end{pmatrix}, \beta = \begin{pmatrix} 1 & 2 & 3 \\ 2 & 1 & 3 \end{pmatrix}, \gamma = \begin{pmatrix} 1 & 2 & 3 \\ 2 & 3 & 1 \end{pmatrix}\right.$$
$$\left.\delta = \begin{pmatrix} 1 & 2 & 3 \\ 3 & 1 & 2 \end{pmatrix}, \epsilon = \begin{pmatrix} 1 & 2 & 3 \\ 3 & 2 & 1 \end{pmatrix}\right\} \tag{5-9}$$

Table 5-4

	1	α	β	γ	δ	ϵ
1	1	α	β	γ	δ	ϵ
α	α	1	γ	β	ϵ	δ
β	β	δ	1	ϵ	α	γ
γ	γ	ϵ	α	δ	1	β
δ	δ	β	ϵ	1	γ	α
ϵ	ϵ	γ	δ	α	β	1

Notice that the group S_3 is non-Abelian since $\alpha\beta \neq \beta\alpha$.

Dirichlet's *Pigeonhole Principle* states that if $n + 1$ objects have to be placed in n locations, then at least two of the objects must end up in the same location. We apply this self-evident principle to obtain another representation of permutations.

Let α be a permutation of the set $X = \{1,2, \ldots, n\}$ and $u \in X$. The sequence

$$u, (u)\alpha, (u)\alpha^2, \ldots, (u)\alpha^n \qquad \textbf{(5-10)}$$

contains $n + 1$ terms, but each of these is an integer between 1 and n. Hence, (5-10) contains a repetition, which means that

$$(u)\alpha^i = (u)\alpha^j$$

for some integers $0 \leq i < j \leq n$. Multiplying both sides of this relation by α^{-i}, $u = (u)\alpha^{j-i}$ where $j - i$ is positive. Now let t be the smallest positive integer for which $u = (u)\alpha^t$. Then each term of the sequence

$$u, (u)\alpha, \ldots, (u)\alpha^{t-1} \qquad \textbf{(5-11)}$$

is distinct. Moreover, if n is any integer, then (5-11) contains $(u)\alpha^n$ for we may write $n = qt + r$ for some quotient q and remainder r, where $0 \leq r < t$. Then,

$$\begin{aligned} (u)\alpha^n &= (u)\alpha^{qt+r} = (u)(\alpha^t)^q\alpha^r \\ &= \underbrace{(((((u)\alpha^t)\alpha^t) \ldots)\alpha^t)\alpha^r}_{q \text{ factors}} \\ &= (u)\alpha^r \qquad \text{where } 0 \leq r \leq t - 1 \end{aligned}$$

Sequence (5-11) is called the *orbit* of u under the permutation α. By what we have just shown, the orbit (5-11) contains $(u)\alpha^n$ for all integers n.

Now if $v \in X$ is not contained in the orbit, then we may form its orbit

$$v, (v)\alpha, \ldots, (v)\alpha^s$$

and the orbits (5-11) and (5-12) have no element in common. For if $(v)\alpha^i = (u)\alpha^j$ for some i,j, then, $v = (u)\alpha^{j-i}$ would belong to the orbit (5-11), contrary to assumption.

In this way, we may partition the set X into orbits with respect to the permutation α.

For example, if

$$\alpha = \begin{pmatrix} 1 & 2 & 3 & 4 & 5 & 6 & 7 & 8 \\ 2 & 8 & 7 & 5 & 6 & 4 & 3 & 1 \end{pmatrix} \qquad \textbf{(5-13)}$$

then, its orbits are

1, 2, 8 (the orbit containing 1)
3, 7 (the orbit containing 3)
4, 5, 6 (the orbit containing 4)

Corresponding to the preceding orbits we form the *cycles*

(128)
(37)
(456)

and write permutation (5-13) as

$$\alpha = (128)\ (37)\ (456) \tag{5-14}$$

The form (5-14) completely describes the permutation α. To find $(i)\alpha$, we look for the cycle containing i. If i is the very last term of that cycle, then $(i)\alpha =$ the first term of the cycle. If i is not the last term of the cycle, then $(i)\alpha$ is the term immediately following i in the cycle.

Observe that as a permutation α of the set $X = \{1,2, \ldots, n\}$ partitions X into its orbits, the corresponding cycles of α will have no element in common. We say that the cycles of α are *disjoint*.

It should be observed that a permutation, when written in the form of disjoint cycles, may have a number of different forms. For example, for the permutation α of (5-14), we may just as well write

$$\alpha = (281)(73)(645)$$

or

$$\alpha = (37)(812)(564)$$

Some other examples of permutations presented in disjoint cycle form are

$$\begin{pmatrix} 1 & 2 & 3 & 4 & 5 & 6 \\ 2 & 1 & 6 & 5 & 3 & 4 \end{pmatrix} = (12)(3645)$$

$$\begin{pmatrix} 1 & 2 & 3 & 4 \\ 2 & 1 & 4 & 3 \end{pmatrix} = (12)(34)$$

$$\begin{pmatrix} 1 & 2 & 3 & 4 & 5 \\ 2 & 4 & 3 & 1 & 5 \end{pmatrix} = (124)(3)(5)$$

The last of these permutations does not move 3 or 5 at all. In this case it is common practice to simply delete their symbols and write

$$\begin{pmatrix} 1 & 2 & 3 & 4 & 5 \\ 2 & 4 & 3 & 1 & 5 \end{pmatrix} = (124)$$

One advantage of writing a permutation α in cycle form is that one may see at a glance other permutations whose product is α. For example, if

$$\alpha = \begin{pmatrix} 1 & 2 & 3 & 4 & 5 \\ 2 & 1 & 4 & 5 & 3 \end{pmatrix}$$

then in cycle form

$$\alpha = (12)(345)$$

However,

$$(12) = \begin{pmatrix} 1 & 2 & 3 & 4 & 5 \\ 2 & 1 & 3 & 4 & 5 \end{pmatrix}$$

and

$$(345) = \begin{pmatrix} 1 & 2 & 3 & 4 & 5 \\ 1 & 2 & 4 & 5 & 3 \end{pmatrix}$$

Thus

$$\begin{pmatrix} 1 & 2 & 3 & 4 & 5 \\ 2 & 1 & 4 & 5 & 3 \end{pmatrix} = \begin{pmatrix} 1 & 2 & 3 & 4 & 5 \\ 2 & 1 & 3 & 4 & 5 \end{pmatrix}\begin{pmatrix} 1 & 2 & 3 & 4 & 5 \\ 1 & 2 & 4 & 5 & 3 \end{pmatrix}$$

A *k-cycle* is one of the form $(a_1a_2 \ldots a_k)$. However a 2-cycle (a_1a_2) is called a *transposition*. It is an important fact that any cycle can be written as a product of transpositions. Explicitly, the k-cycle $(a_1a_2 \ldots a_k) = (a_1a_2)(a_1a_3) \ldots (a_1a_k)$. Thus, for example, $(1543) = (15)(14)(13)$. Notice that each transposition has order 2; for example, $(12)(12) = 1$, i.e., $(12)^{-1} = (12)$.

Proposition 5-4. Every permutation can be written as a product of transpositions.

Proof: Writing permutation α in its cycle form and then expressing each of these cycles as a product of transpositions, we obtain a representation of α as a product of transpositions.

A permutation is called *even* or *odd*, depending on whether it is a product of an even or odd number of transpositions. Now a permutation may be written as a product of transpositions in several ways [e.g., $(12)(14) = (12)(14)(23)(32)$]; so, for the preceding definition to make any sense we must show that no permutation is both even and odd. If in the expression

$$\Delta = \Pi(x_i - x_j)$$
$$1 \leq i < j \leq n$$

we interchange a pair of the subscripts of the x's, then Δ changes sign. Hence, an even permutation of the set $X = \{x_1, x_2, \ldots, x_n\}$ does not change Δ, while an odd permutation of X changes the sign of Δ. Since $\Delta \neq -\Delta$, a permutation cannot be both odd and even.

Proposition 5-5. The set A_n of all even permutations of S_n is a group, called the *alternating group*. The order of A_n is $n!/2$.

Proof: The product of two even permutations is again an even permutation, which is the closure axiom. The associative law is valid throughout S_n; hence, all the more so it is valid in A_n. The identity 1 is contained in A_n since, for example,

$$(12)(12) = \text{the identity}$$

Finally, the inverse of a product of transpositions is the product of these transpositions, written in reverse order [i.e., $((12)(13)(14))^{-1} = (14)(13)(12)$]; hence, the inverse of an even permutation is also an even permutation.

Finally, if the even permutations of S_n are

$$1, \alpha_2, \ldots, a_k \tag{5-15}$$

then the permutations

$$(1,2), \alpha_2(1,2), \ldots, a_k(1,2) \tag{5-16}$$

are all odd. The list (5-16) includes all the odd permutations. For if ρ is odd, then $\rho(1,2)$ is even; whence, $\rho = \rho(1,2)(1,2)$ belongs to the list (5-16). Since then the lists (5-15) and (5-16) contain all the permutations of S_n, and each of (5-15) and (5-16) contain exactly k permutations, it follows that $k =$ the order of $A_n = n!/2$.

5-3 Subgroups, Cosets, and the Theorem of Lagrange

A subgroup of a group G is a subset H satisfying the following properties:

Property 5-1. a,b in $H \Rightarrow ab$ is in H

Property 5-2. a in $H \Rightarrow a^{-1}$ is in H

A subgroup H tacitly contains at least one element a; hence, H contains $1 = aa^{-1}$ because of Properties 5-1 and 5-2. It follows that if H is a subgroup of G, then H is itself a group relative to the given multiplication of G, for the associative law is valid in H by virtue of the fact that it holds throughout the group G.

If the operation of the group G is addition, rather than multiplication, then Properties 5-1 and 5-2 for a subgroup H take a different form. That is, in additive notation, they are the following:

Property 5-3. a, b in $H \Rightarrow a + b$ is in H

Property 5-4. a in $H \Rightarrow -a$ is in H

Example 5-2. In the group I of integers under the operation of addition, the set E of even integers is a subgroup. The symmetric groups contain many subgroups. For example, the alternating group A_n is a subgroup of S_n. Also, in the group S_3 of permutations of the symbols A, B, and C those permutations of S_3 which do not move the symbol A form a subgroup. Equally

$$H = \{\sigma \in S_3 \mid (B)\sigma = B\}$$

is a subgroup. Its elements are $1 = \begin{pmatrix} A & B & C \\ A & B & C \end{pmatrix}$ and $\begin{pmatrix} A & B & C \\ C & B & A \end{pmatrix}$. The set G of nonzero complex numbers form a group under the operation of multiplication and the set H of all complex numbers z of absolute value $|z| = 1$ is a subgroup.

While it is normally straightforward to verify Properties 5-1 and 5-2 for a finite set H, there is even an easier test for determining whether H is a subgroup.

Theorem 5-1. If H is a finite subset of a group G such that

$$a, b \text{ in } H \Rightarrow ab \text{ is in } H$$

then, H is a subgroup.

Proof: Suppose the distinct elements of H are $h_1, \ldots, h_n$, then, for any h_i, the elements $h_1h_i, \ldots, h_nh_i$ belong to H, and they are distinct because of the Cancellation Law. Therefore, the preceding elements are precisely the elements $h_1, \ldots, h_n$ of H in perhaps some other order. In particular, one of these elements, for example h_jh_i, is h_i, $h_jh_i = h_i$. Hence, $h_j = 1$. Since now H is known to contain the identity, one of the elements of $h_1h_i, \ldots, h_nh_i$ must $= 1$. Suppose, for example, $h_kh_i = 1$. Then $h_k = (h_i)^{-1}$. We have shown that H contains the inverse of any of its elements h_i. Therefore, H is a subgroup because of Properties 5-1 and 5-2.

Typically, a subgroup consists of these group elements which have in common a certain property. Can the remaining elements be classified according to some other properties?

The answer is "yes," as we shall now show.

For a subgroup H of a group G and elements x and y of G, put the following:

Property 5-5. $x \backsim y$ if xy^{-1} belongs to H.

We remarked earlier that H contains the identity 1; hence, $x \backsim x$ for every x in G. The relation Property 5-5 is, therefore, reflexive. If $x \backsim y$, then $xy^{-1} \in H$; hence, $(xy^{-1})^{-1} \in H$ because of Property 5-2. However, $(xy^{-1})^{-1} = yx^{-1}$. Thus, $y \backsim x$, i.e., Property 5-5 is also symmetric. Finally, if $x \backsim y$ and $y \backsim z$, then H contains xy^{-1} and yz^{-1}; hence, product $xz^{-1} = (xy^{-1})(yz^{-1})$ is also in H. Therefore, $x \backsim z$. Altogether, the relation (Property 5-5) is an equivalence relation on G and the elements of G are partitioned into distance equivalence classes by this relation. We describe now these equivalence classes.

The *right coset* Hx consists of all elements of the form hx, for some h in H. Cosets are relevant for the following reason:

Property 5-6. $x \backsim y \Leftrightarrow y \in Hx$

Indeed, if y belongs to Hx, then $y = hx$ for some h in H and $yx^{-1} = h$; thus, $y \backsim x$. Reversing this argument shows that y belongs to Hx if $y \backsim x$.

We have proved the following proposition:

> **Proposition 5-6.** If H is a subgroup of a group G, then the elements of G are partitioned into right cosets of H. Moreover, two right cosets are either identical or have no elements in common.

Remark 5-3. Equally, the group G may be partitioned into the *left* cosets xH relative to H, where

$$H = \{g \in G \mid g = xh \quad \text{for some } h \text{ in } H\}$$

If the distinct right cosets of G relative to H are H, Hx, Hy, . . . , then

$$G = H \cup Hx \cup Hy \cup \ldots \tag{5-17}$$

and Equation (5-17) is called the (right) coset decomposition of G relative to H.

Another fundamental property of cosets is the following:

Property 5-7. Hx and Hy are equipotent (i.e., have the same number of elements). Since $H = H \cdot 1$ is also a coset, it follows then that any coset Hx has exactly as many elements as the subgroup H.

To verify Property 5-7, a one-to-one onto map $f: Hx \to Hy$ must be established. This is easy — define

$$f: hx \to hy$$

the mapping f is onto (by definition of Hy) and is one-to-one because of the cancellation law.

Now it is possible to prove the most important theorem in the theory of groups.

Theorem 5-2. *Lagrange's Theorem.* If G is a group of finite order n and H is a subgroup of G of order m, then m divides n.

Proof: Decomposing G into distinct cosets relative to H

$$G = H \cup Hx_2 \cup \ldots \cup Hx_t \qquad \textbf{(5-18)}$$

each of these cosets contains m elements, and as they have no elements in common, $H \cup Hx_2 \cup \ldots \cup Hx_t$ contains mt elements. Hence, $n = mt$, and m divides n.

In Equation (5-18) the number t of distinct cosets is called the *index* of H in G and we write $t = [G : H]$ for the index. Lagrange's Theorem states $[G : 1] = [G : H][H : 1]$.

Remark 5-4. We have denoted the coset containing x by Hx, for a subgroup H. If the group operation is addition, then it is more natural to write $H + x$ for this coset. Using the additive notation, Equation (5-17) reads

$$G = H \cup H + x \cup H + y \cup \ldots$$

5-4 Groups and Geometry

The theory of groups has numerous applications in many branches of mathematics. Geometry was particularly influenced by the study of groups, once it was noticed that some of the more basic concepts of geometry could be formulated in the language of the theory of groups.

Consider, for example, the notion of congruence. Two geometric figures are congruent if it is possible to move one of the figures so that it coincides with the other. (See Figure 5-1, C and C' are congruent.) What is such a movement σ? It is nothing more than a permutation of the points of the Euclidean plane which preserves distances between the points of the plane. The figures C and C' are, therefore, congruent if there exists such a permutation σ which maps C onto C'.

An isometry of the ordinary Euclidean plane is a permutation σ of points of the plane such that for any two points P and Q of the plane the distance between P and Q is the same as the distance between $(P)\sigma$ and $(Q)\sigma$.

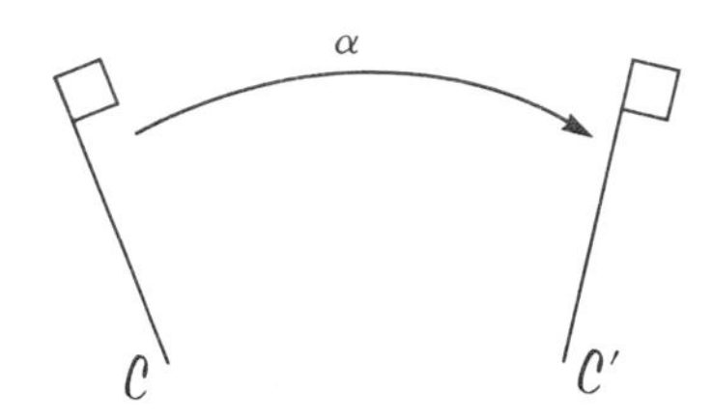

Figure 5-1

Clearly, the product of two isometrics is an isometry and the inverse of an isometry is an isometry. Therefore, the set I of all isometries forms a subgroup of the symmetric group of all permutations of the points of the plane. The group I is called the *isometry group* of the plane. Examples of isometries are rotations, translations, and reflections about a line.

We quote, without proof, the following result:*

Proposition 5-7. The only isometry σ such that $(A)\sigma = A$, $(B)\sigma = B$, and $(C)\sigma = C$ for three noncollinear points A, B, and C is $\sigma = 1$.

Corollary 5-1. If β and γ are isometries such that $(A)\beta = (A)\gamma$, $(B)\beta = (B)\gamma$, and $(C)\beta = (C)\gamma$ for three noncollinear points A, B, and C, then $\beta = \gamma$.

Proof: The permutation $\sigma = \beta\gamma^{-1}$ fixes each of the points A, B, and C; hence, $\sigma = 1 = \beta\gamma^{-1}$. Consequently, $\beta = \gamma$.

---

*See *Theory and Problems of Group Theory*, Baumslag, B., and Chandler, B. Schaum's Outline Series, McGraw-Hill Book Company, 1968.

Corollary 5-1 states that an isometry is completely determined once we know what it does to three noncollinear points. This fact is quite useful in the calculation of isometries.

The degree of symmetry of a geometric figure can be measured by the number of isometries which leave the figure unchanged. If $\mathcal{C}$ is a set of points in the plane, then the set $I(\mathcal{C})$ of all isometries which map $\mathcal{C}$ onto itself is a subgroup of I — this group is called the *isometry group of* $\mathcal{C}$. For example, if $\mathcal{C}$ is the square shown in Figure 5-2, then the 90° rotation σ of the plane about the center of $\mathcal{C}$ belongs to $I(\mathcal{C})$. On the other hand a 60° rotation about the center of $\mathcal{C}$ is an isometry, but it does not belong to $I(\mathcal{C})$.

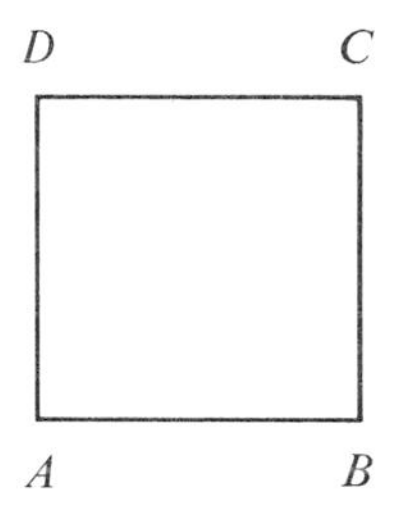

Figure 5-2

Let us calculate the order of $I(\mathcal{C})$ for the square $\mathcal{C}$ of Figure 5-2. The 90° rotation σ belongs to $I(\mathcal{C})$ as do σ^2, σ^3, and $\sigma^4 = 1$. If β is the reflection of $\mathcal{C}$ through the southwest-northeast axis of $\mathcal{C}$, then $I(\mathcal{C})$ contains β;

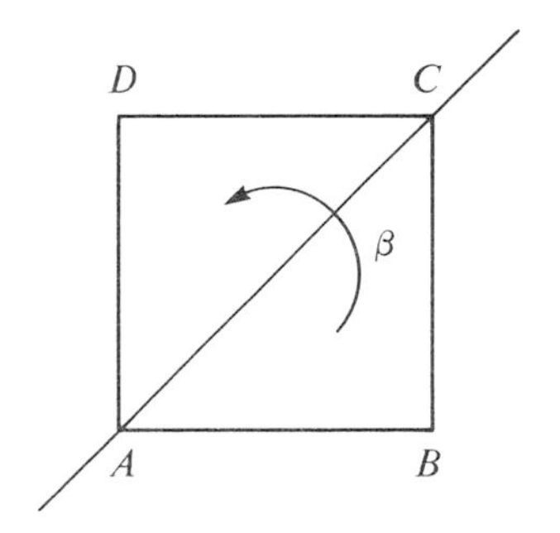

Figure 5-3

hence, $\sigma\beta$, $\sigma^2\beta$, and $\sigma^3\beta$ belong to $I(\mathcal{C})$ as well. (See Figure 5-3.) Therefore, $I(\mathcal{C})$ contains the isometries

$$1,\ \sigma,\ \sigma^2,\ \sigma^3,\ \beta,\ \sigma\beta,\ \sigma^2\beta,\ \sigma^3\beta \tag{5-19}$$

and what is more, it can be verified that the isometries (5-19) are all different. Hence, the order of $I(\mathcal{C})$ is at least eight.

Now the most distant pairs of points of $\mathcal{C}$ are (A,C) and (B,D). Hence, if $\gamma \in I(\mathcal{C})$, then on the vertices

$$\gamma = \begin{pmatrix} A & C \\ C & A \end{pmatrix} \tag{5-20}$$

or

$$\gamma = \begin{pmatrix} A & C \\ A & C \end{pmatrix} \tag{5-21}$$

or

$$\gamma = \begin{pmatrix} A & C \\ B & D \end{pmatrix} \tag{5-22}$$

or

$$\gamma = \begin{pmatrix} A & C \\ D & B \end{pmatrix} \tag{5-23}$$

If γ is as in (5-20), i.e., if γ interchanges A and C, then either (1) γ fixes B and D or (2) γ interchanges B and D (because γ preserves distances between the vertices of $\mathcal{C}$). In case (1) $\gamma = \begin{pmatrix} A & B & C \\ C & B & A \end{pmatrix}$. However, note Figure 5-4. Hence, on the vertices, $\sigma^2\beta = \begin{pmatrix} A & B & C \\ C & B & A \end{pmatrix}$. Since the isometries γ and $\sigma^2\beta$ move the vertices, A, B, and C in the same way, it follows from Corollary 5-1 that $\gamma = \alpha^2\beta$. By a similar analysis it may be shown in case (2) and (5-20), (5-21), (5-22), and (5-23) that γ is one of the isometries (5-19). Therefore, $I(\mathcal{C})$ has order 8.

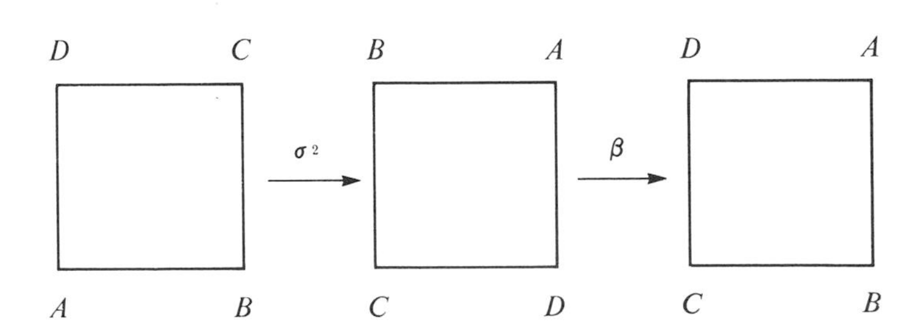

Figure 5-4

Some geometric configurations may be expressed in terms of permutations. The triangles ABC and $A'B'C'$ in Figure 5-5 are said to be *perspective* from the point P. While in Figure 5-6, ABC and $A'B'C'$ are said to be *perspective from infinity* if the lines AA', BB' and CC' are parallel.

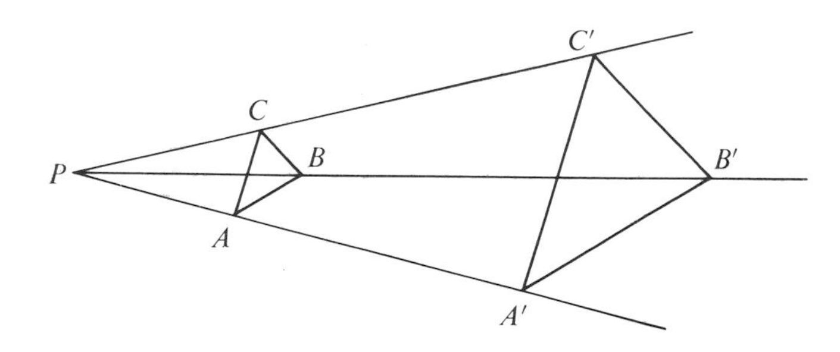

Figure 5-5

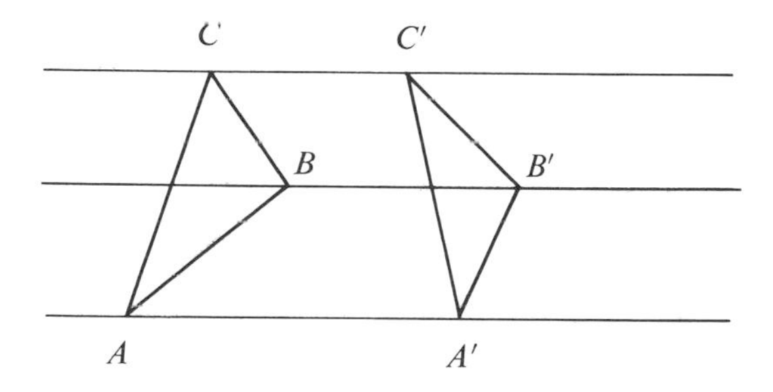

Figure 5-6

Desargue's Theorem in geometry states: If ABC and $A'B'C'$ are triangles which are perspective from infinity such that AB is parallel to $A'B'$ and BC is parallel to $B'C'$, then AC is parallel to $A'C'$. See Figure 5-7. Figure 5-7 is sometimes called *Desargue's configuration.*

In terms of permutations, Desargue's Theorem can be derived from the following fact: Given any pair of points B and B' of the plane, there is a translation T of the plane which maps B onto B'.

Indeed, to prove Desargue's Theorem, we only need to observe that a translation always moves a line l onto a line l' which is parallel to l. Thus, letting T be a translation moving B to B', the line joining B' and $T(C)$

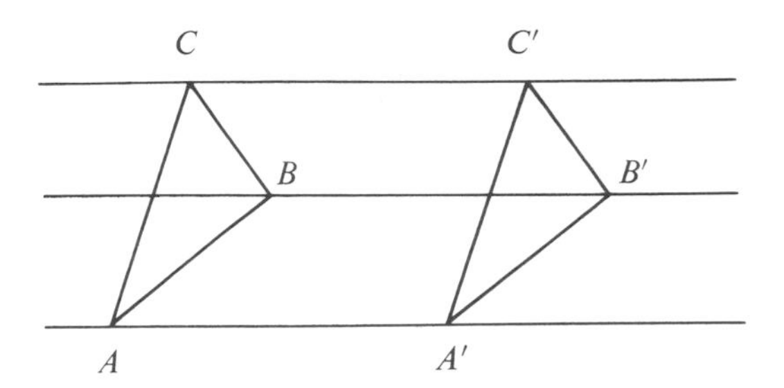

Figure 5-7

must be parallel to the line joining B and C. Hence, $T(C) = C'$. Similarly, T maps A onto A'. Therefore, as T maps AC onto $A'C'$, AC and $A'C'$ must be parallel.

The importance of Desargue's configuration to the general theory of affine geometries will be explained in Section 5-14.

A *collineation* of an affine geometry Π is a permutation σ of the points of Π such that whenever A, B, and C are collinear points of Π, so are $(A)\sigma$, $(B)\sigma$, and $(C)\sigma$ collinear. Thus, a collineation, besides mapping points onto points, also maps lines onto lines.

The set G of all collineations of an affine geometry Π is a subgroup of the symmetric group of all permutations of the points of Π, and G is called the *collineation group* of Π. The order of the collineation group G may be quite large, even when Π contains only a few points (see Section 5-7). Some important groups may be realized as collineation groups of affine geometrics.

5-5 Cyclic Groups and the Order of an Element

If G is a group of finite order and $a \neq 1$ is an element of G, then the powers of a

$$a, a^2, a^3, \ldots \tag{5-24}$$

cannot all be different. Hence, for some integers r and s, $r > s$, $a^r = a^s$, and after canceling, $a^{r-s} = 1$. The *order* of a is defined to be the smallest positive integer m for which $a^m = 1$.

Theorem 5-3. If G is a group of finite order and $a \neq 1$ is an element of G of order m, then: (a) $H = \{1, a, \ldots, a^{m-1}\}$ is a subgroup of order m of G; (b) m divides the order of G; (c) H contains every power a^n of a. (d) If $a^k = 1$ for an integer k, then m divides k.

Proof: The set H of all power of a is a subgroup, for if a^n and a^m are powers of a so is $a^n a^m = a^{n+m}$ a power of a. For any power a^n of a, we may divide n by m to get a quotient and remainder, i.e.,

$$n = qm + r$$

where q and r are integers and $0 \leq r < m$, then $a^n = a^{qm+r} = a^{qm}a^r = (a^m)^q a^r = a^r$. Thus, every power a^n of a is actually a^r, $0 \leq r < m$. In particular the elements of H are contained in the set

$$\{1, a, \ldots, a^{m-1}\} \quad \textbf{(5-25)}$$

From the definition of order, it is immediate that the elements in the set (5-25) are distinct. Hence, $H = \{1, a, \ldots, a^{m-1}\}$, which proves (a) and (c) of Theorem 5-3. Moreover, part (b) of Theorem 5-3 follows immediately from Lagrange's Theorem.

Finally, if $a^k = 1$ for some integer k, then applying the division algorithm we have

$$k = qm + r$$

for some quotient q and remainder r, $0 \leq r < m$. Then $1 = a^k = (a^m)^q a^r = a^r$. However, as m is the order of a, the integer r cannot be positive. Therefore, $r = 0$ and m divides k.

Corollary 5-2. If G is a group of finite order n, then $a^n = 1$ for every a in G.

Proof: Each $a = 1$ has an order m which divides n. Therefore $a^n = (a^m)^{n/m} = 1^{n/m} = 1$.

A group finite G is called *cyclic* if it contains an element a such that every element of G is a power of a. The element a is called a *generator* of G.

An example of a cyclic group of order n is the group $I/(n)$ under the operation of addition. For $[1] + [1] = [2]$, $[1] + [1] + [1] = [3], \ldots$. So $[1]$ is a generator of G.

It is an immediate consequence of Theorem 5-3 that if G is a cyclic group of order n and if a is a generator of G, then a has order n.

It is also true that if G and G' are cyclic groups of order n, then they are isomorphic. For if a is a generator of G and b is a generator of G', then $G = \{1,a,a^2, \ldots, a^{n-1}\}$ and $G' = \{1,b,b^2, \ldots, b^{n-1}\}$ have operation Tables 5-5 and 5-6.

Table 5-5

	1	1	a^2	$\cdots$	a^{n-1}
1	1	a	a^2	$\cdots$	a^{n-1}
a	a	a^2	a^3	$\cdots$	1
a^2	a^2	a^3	a^4	$\cdots$	a
$\vdots$	$\vdots$	$\vdots$			$\vdots$

Table 5-6

	1	b	b^2	$\cdots$	b^{n-1}
1	1	b	b^2	$\cdots$	b^{n-1}
b	b	b^2	b^3	$\cdots$	1
b^2	b^2	b^3	b^4	$\cdots$	b
$\vdots$	$\vdots$	$\vdots$			$\vdots$

Clearly, the mapping $a^i \to b^i$ defines an isomorphism between G and G'.

We remarked before that the group $I/(n)$ is cyclic of order n. Therefore, we have the following theorem:

Theorem 5-4. Any finite cyclic group G of order n is isomorphic to $I/(n)$.

The notion of cyclic group can be extended to infinite groups. Indeed, any group G (infinite or not) is called *cyclic* if each element of G is a power of some fixed element of G.

There is only one infinite cyclic group.

Theorem 5-5. If G is an infinite cyclic group, then G is isomorphic to the group of integers.

Proof: Let a be a generator of G. Define a mapping $f: I \longrightarrow G$ by the rule

$$f(n) = a^n$$

for an integer n. Then f is onto, because a is a generator G. Moreover, f is one-to-one. For if $f(n) = f(m)$, then $a^{n-m} = 1$. If $n - m \neq 0$, then a has some finite order. But then a could not generate the infinite group G (see Theorem 5-3). Therefore, $n = m$, which implies f is one-to-one.

Finally, for integers n and m,

$$f(n + m) = a^{n+m} = a^n a^m = f(m)f(n)$$

consequently, f is an isomorphism.

Every group G has at least two subgroups. For $\{1\}$, the set consisting of the identity element is a subgroup (or order 1). Also G is a subgroup of itself. Generally, a subgroup H of G is called a *nontrivial* subgroup if $H = \{1\}$ and $H \neq \{G\}$.

Practically all groups have nontrivial subgroups, as shown by the next result.

Theorem 5-6. The only groups G which do not have a nontrivial subgroup are the cyclic groups of prime order and the group $G = \{1\}$ of order one.

Proof: If $G \neq \{1\}$, then G has an element $a \neq 1$. The set

$$H = \{a^n \mid n \text{ is an integer}\}$$

consisting of all the powers of a is a subgroup $\neq \{1\}$. Hence, if G has no nontrivial subgroup, then $H = G$, which implies G is cyclic. If G is infinite, then G is isomorphic to I, the group of integers under addition. But I contains the subgroup E consisting of all even integers. Therefore, we may assume that G has finite order m, whence, $m =$ the order of a. If m is not prime, then m factors $m = pq$ as a product of integers $1 < p < m$, $1 < q < m$. Put $b = a^p$, then $b^q = 1$; hence, b has some order $t \leq p$. The set

$$K = \{1, b, \ldots, b^{t-1}\}$$

is a subgroup of G such that $K \neq \{1\}$ and $K \neq G$. This shows the group must have prime order.

5-6 Application of Groups to Number Theory

Lagrange's Theorem showed that number theoretic considerations enter quite naturally in the study of finite groups. Conversely, using the theory of groups, we get results in the theory of numbers.

For a positive integer n let G_n denote the set of equivalence classes $[a]$ modulo n where $(a,n) = 1$. Now, if $[a]$ and $[b]$ belong to G_n so does their product $[a][b]$. For $(a,n) = (b,n) = 1$ implies $(ab,n) = 1$; hence, $[a][b] = [ab] \in G_n$. Moreover, if $[a] \in G_n$, then

$$1 = ax + ny$$

for some integers x and y. So, $[a][x] = [1]$. That is, $[a]^{-1}$ exists and is an element of G_n. It follows that G_n is a group under the product rule for equivalence classes, and the order of G_n is $\phi(n)$, where ϕ is the Euler ϕ-function.

Theorem 5-7. *Fermat's Theorem.* If $(a,n) = 1$, then $a^{\phi(n)} \equiv$ (mod n).

Proof: The equivalence class $[a]$ belongs to the group G_n, which is of order $\phi(m)$. Therefore, $a^{\phi(m)} = [1]$ because of Corollary 5-2. Reading this last relation as a congruence, $a^{\phi(m)} \equiv 1 \pmod{n}$.

Let us now consider the group G_p, where p is a prime. If $[a]$ is any element of G_p; then $(a,p) = 1$ and dividing p into a to get a remainder, we have $[a] = [r]$ where $0 \leq r \leq p - 1$. Therefore, we may take for the elements of G the equivalence classes $[a]$, where $0 \leq a \leq p - 1$. Now, if $[a]$ is of order 2 in the group G_p, then $[a]^2 = [1]$ or $a^2 - 1 \equiv 0 \pmod{p}$. Therefore, $(a - 1)(a + 1) \equiv 0 \pmod{p}$. Since $a < p$, $a - 1 \not\equiv 0$ (mod p). Therefore, $a + 1 \equiv 0 \pmod{p}$. But because of the size of a, this implies $a = p - 1$. Hence, of the elements

$$[1], [2], \ldots, [p - 1]$$

of G_p, only $[p - 1]$ is of order 2.

Now if G is any Abelian group of finite order and if

$$a_1, a_2, \ldots, a_r$$

are the elements of G which are not of order two, then their product

$$a_1 a_2 \ldots a_r = 1$$

Indeed, since a_1 has order $\neq 2$, its inverse will be one of the elements a_i, $i \neq 1$. In the product $a_1 \ldots a_r = a_1 \ldots a_{i-1} a_i a_{i+1} \ldots a_r$, we may bring to the left the factor a_i (because G is Abelian), and write

$$a_1 \ldots a_{i-1} a_i a_{i+1} \ldots a_r = (a_1 a_i) \ldots a_{i-1} a_{i+1} \ldots a_r$$

However, $a_1 a_i = 1$. By similarly canceling a_2 with its inverse, etc, it is evident that the product $a_1 a_2 \ldots a_r$ collapses to $a_1 a_2 \ldots a_r = 1$.

We may apply this interesting fact to the group G_p. The sequence

$$[1], [2], \ldots, [p-2]$$

consists of precisely those elements of G_p whose order is different from 2. Therefore,

$$\begin{aligned}[1][2] \ldots [p-2][p-1] &= ([1][2] \ldots [p-2])[p-1] \\ &= [p-1]\end{aligned}$$

Hence $[1 \cdot 2 \ldots (p-1)] = [p-1]$, or $[(p-1)!] = [-1]$. In terms of congruences,

$$(p-1)! \equiv -1 \pmod{p}$$

We have proved the following theorem:

Theorem 5-8. *Wilson's Theorem.* If p is a prime, then, $(p-1)! \equiv -1 \pmod{p}$.

5-7 Computations with Subgroups

The problem of describing a group is often solved by searching the group for a subgroup and then decomposing the group into cosets relative to the subgroup. In this section two examples of such a procedure will be given.

Example 5-3. Determine all groups G of order 6.

Solution: The order of each element different from the identity in such a group G must be 2, 3, or 6. If G has an element of order 6, then G is isomorphic to the cyclic group $I/(6)$.

Discarding that possibility, each element $\neq 1$ of G has order 2 or 3. Moreover, G has to have an element a of order 3 (see Exercise 5-5, Problem 7). Then,

$$H = \{1,a,a^2\}$$

is a subgroup of G and $G = H \cup Hb$ for some $b \in H$. The elements of G are $\{1,a,a^2,b,ab,a^2b\}$. Of course, ba belongs to G; hence, to one of the cosets H or Hb. If $ba \in H$, then by cancellation, $b \in H$, which is impossible. Thus, $ba = b$, ab, or a^2b. However, $ba = b$ implies $a = 1$, which is again impossible. Thus, either

$$ba = ab \tag{5-26}$$

or

$$ba = a^2b \tag{5-27}$$

A similar analysis may be applied to b^2 to show that $b^2 \in H$. Thus,

$$b^2 = 1 \tag{5-28}$$

or

$$b^2 = a \tag{5-29}$$

or

$$b^2 = a^2 \tag{5-30}$$

In case (5-29) $b^3 = ab \neq 1$ and b is necessarily of order 6. Similarly in (5-30), $b^3 = a^2b \neq 1$ and b is of order 6. Having earlier ruled out the case where G is cyclic, it follows that

$$b^2 = 1$$

Now if $ab = ba$, then $(ab)^2 = a^2b^2 = a^2 \neq 1$ and $(ab)^3 = a^3b^3 = b \neq 1$; whence, ab is of order 6. Therefore, $ba = a^2b$.

Altogether, if G is a group of order 6, then either G is cyclic or $G = H \cup Hb$ where $H = \{1,a,a^2\}$, $a^3 = b^2 = 1$ and $ba = a^2b$. Using those relations, we may compute the operation table for G in the case G is not cyclic. See Table 5-7.

Example 5-4. From the field $I/(3)$ an affine geometry Π of 9 points and 12 lines may be constructed. The points of this geometry are $A = (0,0)$,

Table 5-7

	1	a	a^2	b	ab	a^2b
1	1	a	a^2	b	ab	a^2b
a	a	a^2	1	ab	a^2b	b
a^2	a^2	1	a	a^2b	b	ab
b	b	a^2b	ab	1	a^2	a
ab	ab	b	a^2b	a	1	a^2
a^2b	a^2b	ab	b	a^2	a	1

B – (1,0), $C = (2,0)$, D = (0,1), $E = (1,1)$, $F = (2,1)$, $G = (0,2)$, $H =$ (1,2), $I = (2,2)$, and the lines are

$$
\begin{array}{llll}
l_1 = \{A,B,C\} & l_2 = \{D,E,F\} & l_3 = \{G,H,I\} & l_4 = \{A,D,G\} \\
l_5 = \{B,E,H\} & l_6 = \{C,F,I\} & l_7 = \{A,E,I\} & l_8 = \{C,D,H\} \\
l_9 = \{B,F,G\} & l_{10} = \{A,F,H\} & l_{11} = \}B,D,I\} & l_{12} = \{C,E,G\}
\end{array}
$$

The 12 lines of this geometry fall into 4 classes, each consisting of 3 parallel lines. Generally a set of parallel lines is called a *pencil*. Thus, the pencils of this geometry are

$$
\begin{array}{ll}
P_1 = \{l_1, l_2, l_3\} & P_2 = \{l_4, l_5, l_6\} \\
P_3 = \{l_7, l_8, l_9\} & P_4 = \{l_{10}, l_{11}, l_{12}\}
\end{array}
$$

Recall that a collineation of Π is a permutation of the points of Π which maps the lines of Π onto lines (see Section 5-4). For example

$$
\sigma = \begin{pmatrix} A & B & C & D & E & F & G & H & I \\ C & F & I & B & E & H & A & D & G \end{pmatrix} \quad \textbf{(5-31)}
$$

is a collineation, for

$$\sigma \begin{cases} l_1 \to l_6 \\ l_2 \to l_5 \\ l_3 \to l_4 \\ l_4 \to l_1 \\ l_5 \to l_2 \\ l_6 \to l_3 \\ l_7 \to l_{12} \\ l_8 \to l_{11} \\ l_9 \to l_{10} \\ l_{10} \to l_8 \\ l_{11} \to l_9 \\ l_{12} \to l_7 \end{cases}$$

As a permutation of the lines of Π

$$\sigma = \begin{pmatrix} l_1 & l_2 & l_3 & l_4 & l_5 & l_6 & l_7 & l_8 & l_9 & l_{10} & l_{11} & l_{12} \\ l_6 & l_5 & l_4 & l_1 & l_2 & l_3 & l_{12} & l_{11} & l_{10} & l_8 & l_9 & l_7 \end{pmatrix} \quad \textbf{(5-32)}$$

On the other hand, the permutation

$$\mu = \begin{pmatrix} A & B & C & D & E & F & G & H & I \\ B & C & D & E & F & G & H & I & A \end{pmatrix}$$

is not a collineation because μ maps the line $l_1 = \{A,B,C\}$ onto $\{B,C,D\}$, which is not a line.

Let us calculate the order of the group G of all collineations of this geometry.

At first glance this may appear to be a hopeless task, for there are altogether 9! permutations of the points of Π, and of these we must find those which preserve the lines of Π. The key to the solution will be to decompose the group G into cosets relative to a subgroup. To see how the different cosets are found, it is convenient to list the points of Π on a grid, depending upon their coordinates. See Figure 5-8. It will also be useful to observe that since a collineation maps parallel lines onto parallel lines, it must map a pencil of parallel lines onto a pencil of parallel lines, i.e., a collineation permutes the pencils. For example, from the form (5-32) of the permutation (5-31) we see that

$$\sigma \begin{cases} P_1 \to P_2 \\ P_2 \to P_1 \\ P_3 \to P_4 \\ P_4 \to P_3 \end{cases}$$

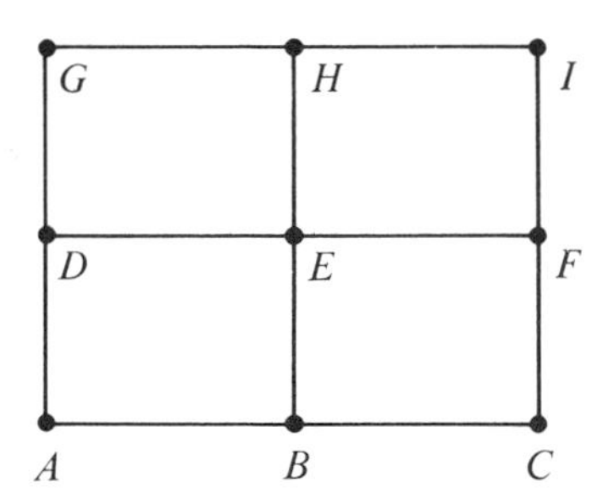

Figure 5-8

or

$$\sigma = \begin{pmatrix} P_1 & P_2 & P_3 & P_4 \\ P_2 & P_1 & P_4 & P_3 \end{pmatrix}$$

as a permutation of the pencils P_1, P_2, P_3, P_4.

Let $H = \{\sigma \in G \mid \sigma : P_1 \rightarrow P_1\}$. The set H consists of those collineations which leave the pencil P_1 unchanged. That is, a collineation in H permutes the lines l_1, l_2, l_3 amongst themselves. Certainly the product of two collineations of H is itself in H; hence, H is a subgroup of G.

To decompose G into cosets relative to the subgroup H, it is necessary to find collineations which do not belong to H. One such collineation is

$$\beta = \begin{pmatrix} A & B & C & D & E & F & G & H & I \\ A & E & I & F & G & B & H & C & D \end{pmatrix}$$

(β is found by "rotating" the line $\{ABC\}$ into the line $\{AEI\}$ about the point A — see Figure 5-8.) One may check by direct calculation that β is a collineation. As a permutation of the lines,

$$\beta = \begin{pmatrix} l_1 & l_2 & l_3 & l_4 & l_5 & l_6 & l_7 & l_8 & l_9 & l_{10} & l_{11} & l_{12} \\ l_7 & l_9 & l_8 & l_{10} & l_{12} & l_{11} & l_4 & l_6 & l_5 & l_1 & l_2 & l_3 \end{pmatrix}$$

As a permutation on the pencils,

$$\beta = \begin{pmatrix} P_1 & P_2 & P_3 & P_4 \\ P_3 & P_4 & P_2 & P_1 \end{pmatrix}$$

It is clear that $\beta \notin H$, for $\beta : P_1 \rightarrow P_3$. Moreover,

$$\begin{aligned} \beta^2 &= \begin{pmatrix} P_1 & P_2 & P_3 & P_4 \\ P_2 & P_1 & P_4 & P_3 \end{pmatrix} \\ \beta^3 &= \begin{pmatrix} P_1 & P_2 & P_3 & P_4 \\ P_4 & P_3 & P_1 & P_2 \end{pmatrix} \\ \beta^4 &= \begin{pmatrix} P_1 & P_2 & P_3 & P_4 \\ P_1 & P_2 & P_3 & P_4 \end{pmatrix} \end{aligned} \tag{5-33}$$

Therefore, $\beta^4 \in H$. From (5-33) it follows that H, $H\beta$, $H\beta^2$, $H\beta^3$ are distinct cosets. For

$$\begin{aligned} \sigma \in H &\Rightarrow \sigma : P_1 \rightarrow P_1 \\ \sigma \in H\beta &\Rightarrow \sigma : P_1 \rightarrow P_3 \\ \sigma \in H\beta^2 &\Rightarrow \sigma : P_1 \rightarrow P_2 \\ \sigma \in H\beta^3 &\Rightarrow \sigma : P_1 \rightarrow P_4 \end{aligned} \tag{5-34}$$

Moreover,

$$G = H \cup H\beta \cup H\beta^2 \cup H\beta^3 \tag{5-35}$$

As we remarked earlier, a collineation τ must map the pencil P_1 onto a pencil. If, for example, $\tau : P_1 \rightarrow P_2$, then from (5-33) we have that $\tau(\beta^2)^{-1} : P_1 \rightarrow P_1$. Hence, $\tau(\beta^2)^{-1} \in H$ and $\tau \in H\beta^2$. Equally, $\tau \in H\beta$ if $\tau : P_1 \rightarrow P_3$; $\tau \in H\beta^3$ if $\tau : P_1 \rightarrow P_4$, and $\tau \in H$ if $\tau : P_1 \rightarrow P_1$.

We have from (5-35) that

$$[G : 1] = 4[H : 1] \tag{5-36}$$

Let us now investigate the subgroup H. Put $K = \{\sigma \in H \mid \sigma : l_1 \rightarrow l_1\}$. To decompose H into cosets relative to the subgroup K, we search for a collineation γ such that $\gamma \in H$, $\gamma \notin K$. Looking at Figure 5-8 we see that the pencil P_1 consists of the 3 horizontal lines of that grid. We choose γ to be that collineation which "raises" each of these horizontal lines, i.e.,

$$\gamma = \begin{pmatrix} A & B & C & D & E & F & G & H & I \\ D & E & F & G & H & I & A & B & C \end{pmatrix}$$

Verifying that γ is indeed a collineation, we find

$$\gamma = \begin{pmatrix} l_1 & l_2 & l_3 & l_4 & l_5 & l_6 & l_7 & l_8 & l_9 & l_{10} & l_{11} & l_{12} \\ l_2 & l_3 & l_1 & l_4 & l_5 & l_6 & l_8 & l_9 & l_7 & l_{11} & l_{12} & l_{10} \end{pmatrix}$$

from which it is also evident that $\gamma \in H$.

The cosets K, $K\gamma$, $K\gamma^2$ are characterized as follows:

$$\begin{aligned} K &= \{\sigma \in H \mid \sigma : l_1 \rightarrow l_1\} \\ K\gamma &= \{\sigma \in H \mid \sigma : l_1 \rightarrow l_2\} \\ K\gamma^2 &= \{\sigma \in H \mid \sigma : l_1 \rightarrow l_3\} \end{aligned}$$

Since any collineation belonging to H must map l_1 onto l_1, l_2, or l_3, it is evident that these cosets exhaust H. Thus,

$$H = K \cup K\gamma \cup K\gamma^2 \tag{5-37}$$

and $[H:1] = 3[K:1]$. Thus, from Equation (5-36)

$$[G:1] = 12[K:1] \tag{5-38}$$

Next we decompose K by noting that $L = \{\sigma \in K \mid \sigma : A \to A\}$ is a subgroup of K. The permutation

$$\delta = \begin{pmatrix} A & B & C & D & E & F & G & H & I \\ B & C & A & E & F & D & H & I & G \end{pmatrix}$$

belongs to K, but not to L since it actually moves A.

The cosets L, $L\delta$, $L\delta^2$ may be described as follows:

$$\begin{aligned} L &= \{\sigma \in K \mid \sigma : A \to A\} \\ L\delta &= \{\sigma \in K \mid \sigma : A \to B\} \\ L\delta^2 &= \{\sigma \in K \mid \sigma : A \to C\} \end{aligned} \tag{5-39}$$

Any collineation of K fixes the line l_1; hence, must map the point A onto A, B, or C. Hence, the cosets (5-39) contain all the elements of K, and

$$K = L \cup L\delta \cup L\delta^2$$

thus, $[K:1] = 3[L:1]$, and from Equation (5-38)

$$[G:1] = 36[L:1] \tag{5-40}$$

The set $M = \{\sigma \in L \mid (A)\sigma = A, (B)\sigma = B, (c)\sigma = C\}$ is a subgroup of L and

$$\rho = \begin{pmatrix} A & B & C & D & E & F & G & H & I \\ A & C & B & D & F & E & G & I & H \end{pmatrix}$$

belongs to L, but not to M. Moreover,

$$L = M \cup M\rho$$

so $[L:1] = 2[M:1]$ and

$$[G:1] = 72[M:1] \tag{5-41}$$

The problem of determining the elements of M is not difficult. For if $\lambda \in M$, then

$$\lambda = \begin{pmatrix} A & B & C & D & E & F & G & H & I \\ A & B & C & . & . & . & . & . & . \end{pmatrix}$$

Once $(D)\lambda$ is known, the remaining values $(E)\lambda$, $(F)\lambda$, $(G)\lambda$, $(H)\lambda$, and $(I)\lambda$ will also be known. For example, suppose $(D)\lambda = E$. Then, λ must map $l_4 = \{A,D,G\}$ onto a line. But $\lambda : \{A,D,G\} \to \{A,E,(G)\lambda\}$. For

$\{A,E,(G)\lambda\}$ to be a line, $(G)\lambda = I$, etc. By similar calculations, one finds that the elements of M are

$$1, \begin{pmatrix} A & B & C & D & E & F & G & H & I \\ A & B & C & E & F & D & I & G & H \end{pmatrix}, \begin{pmatrix} A & B & C & D & E & F & G & H & I \\ A & B & C & F & D & E & H & I & G \end{pmatrix},$$
$$\begin{pmatrix} A & B & C & D & E & F & G & H & I \\ A & B & C & G & H & I & D & E & F \end{pmatrix}, \begin{pmatrix} A & B & C & D & E & F & G & H & I \\ A & B & C & H & I & G & F & D & E \end{pmatrix},$$
$$\text{and } \begin{pmatrix} A & B & C & D & E & F & G & H & I \\ A & B & C & I & G & H & E & F & D \end{pmatrix}$$

Thus, $[M : 1] = 4$, and from Equation (5-41), $[G : 1] = 432$.

5-8 Homomorphisms

It is by the notion of *isomorphism* that groups are considered identical. However, some groups, while not isomorphic, are still similar in one respect or other. For example, the group I of integers and the group $I/(n)$, where n is a positive integer, have in common the property of being cyclic. Of course, I and $I/(n)$ are not isomorphic, since they are not of the same order.

Similarity of groups is expressed in terms of homomorphism. If G and G' are groups, then a *homomorphism* is an onto mapping $f : G \to G'$ such that

$$f(ab) = f(a)f(b) \tag{5-42}$$

for all a,b in G. The group G' is called *homomorphic image* of G. A homomorphism f differs from an isomorphism by the fact that f need not be one to one.

If $f : G \to G'$ is a homomorphism, then all information about the group G' may be determined from the group G and the homomorphism f. For the calculation of products in G' may be accomplished by calculating them in G and then transferring the results to G' via the mapping f, as is stated in Equation (5-42).

On the other hand, in a homomorphism $f : G \to G'$ the group G' does not completely determine G. Indeed, if $G' = \{1\}$ is the group of order one and G is any group whatsoever, then the mapping

$$f(a) = 1 \qquad a \text{ in } G$$

defines a homomorphism $f : G \to G'$.

The exponential law $a^m a^n = a^{m+n}$ has an interesting interpretation in terms of homomorphisms. It may be used to prove the following:

> Every group G contains a subgroup which is a homomorphic image of the integers.

For if a belongs to G and

$$H = \{a^n \mid n \text{ is an integer}\}$$

is the subgroup consisting of all powers of a, then the mapping $f: I \to H$ defined by

$$f(n) = a^n$$

for an integer n, is a homomorphism, since $f(m + n) = a^{m+n} = a^m a^n = f(m)f(n)$ for all integers m and n.

Often a homorphism of a group may be found by considering the group from a different point of view.

Example 5-5. The following permutations of A,B,C,D form a group G of order 4:

$$1 = \begin{pmatrix} A & B & C & D \\ A & B & C & D \end{pmatrix} \qquad \alpha = \begin{pmatrix} A & B & C & D \\ B & A & C & D \end{pmatrix}$$

$$\beta = \begin{pmatrix} A & B & C & D \\ B & A & D & C \end{pmatrix} \qquad \gamma = \begin{pmatrix} A & B & C & D \\ A & B & D & C \end{pmatrix}$$

On the other hand, each of the preceding permutations of the four symbols A, B, C, D is a permutation of the symbols A and B alone. For example, α as a permutation of just A and B, is the permutation

$$\bar{\alpha} = \begin{pmatrix} A & B \\ B & A \end{pmatrix}$$

Similarly,

$$\bar{\beta} = \begin{pmatrix} A & B \\ B & A \end{pmatrix}$$

$$\bar{\gamma} = \begin{pmatrix} A & B \\ A & B \end{pmatrix}$$

$$\bar{1} = \begin{pmatrix} A & B \\ A & B \end{pmatrix}$$

Therefore, G can also be thought of as a permutation group on the symbols A and B. Expressing this in the language of homomorphisms, the mapping f defined by

$$f(1) = \bar{1} \qquad f(\alpha) = \bar{\alpha}$$
$$f(\beta) = \bar{\beta} \qquad f(\gamma) = \bar{\gamma}$$

is a homomorphism

$$f: G \to G'$$

where $G' = \left\{\begin{pmatrix} A & B \\ A & B \end{pmatrix}, \begin{pmatrix} A & B \\ B & A \end{pmatrix}\right\}$. The homomorphic image G' has order 2.

Similarly, the mapping g, where

$$g(1) = g(\alpha) = \begin{pmatrix} C & D \\ C & D \end{pmatrix}$$

and

$$g(\beta) = g(\gamma) = \begin{pmatrix} C & D \\ D & C \end{pmatrix}$$

defines a homomorphism $g : G \to G''$, where

$$G'' = \left\{\begin{pmatrix} C & D \\ C & D \end{pmatrix}, \begin{pmatrix} C & D \\ D & C \end{pmatrix}\right\}$$

As may be seen from this example, a homomorphism $f: G \to G'$ of groups need not be an isomorphism.

Proposition 5-8. Let $f: G \to G'$ be a homomorphism, then $f(1)$ is the identity of G' and $f(a)^{-1} = f(a^{-1})$ for every a in G.

Proof: Since f maps G into G', if u is any element of G'; then, $u = f(b)$ for some b in G. Then $f(1)u = f(1)f(b) = f(1 \cdot b) = f(b) = u$. Thus, $f(1)u = u$. The cancellation law for G' implies $f(1)$ is the identity element of G'.

For an element a of G, $f(1) = f(aa^{-1}) = f(a)f(a^{-1})$. Thus, $f(a^{-1})$ is the inverse of $f(a)$, i.e., $f(a^{-1}) = f(a)^{-1}$.

The *kernel* K of a homomorphism $f: G \to G'$ is defined as $K = \{a \in G \mid f(a) = 1, \text{ the identity of } G'\}$.

For the mapping f of Example 5-5, the kernel $= \{1, \gamma\}$, while the kernel of the homomorphism g of the same example consists of the elements 1 and α.

Proposition 5-8 guarantees that the kernel of a homomorphism $f: G \to G'$ contains at least one element, namely the identity of G.

It is the kernel of a homomorphism $f: G \to G'$ which measures to what extent the homomorphism f fails to be an isomorphism.

Proposition 5-9. Let K be the kernel of the homomorphism $f: G \to G'$. Then: (a) K is a subgroup of G and (b) f is an isomorphism $\Leftrightarrow$ $K = \{1\}$.

Proof: If a and b belong to K, then $f(a) = f(b) = 1$. Thus, $f(ab) = f(a)f(b) = 1$, i.e., ab also belongs to K. Moreover, if $f(a) = 1$; then, $f(a^{-1}) = f(a)^{-1} = 1$. Thus, K contains a^{-1} whenever it contains a. This proves (a).

If f is an isomorphism and a belongs to K, then $f(a) = 1$, as well as $f(1) = 1$. Since f is one-to-one, $a = 1$, whence, $K = \{1\}$.

Conversely, suppose $f: G \to G'$ is a homomorphism whose kernel K consists just of the identity element. Then if $f(a) = f(b)$, we have $1 = f(a)f(b)^{-1} = f(a)f(b^{-1}) = f(ab^{-1})$. Thus, ab belongs to K, and $ab^{-1} = 1$. Therefore, $a = b$, which shows f to be one-to-one, hence, an isomorphism.

It is instructive to decompose a group G into cosets relative to the kernel K of a homomorphism $f: G \to G'$. For we have the following result:

$$f(a) = f(b) \Leftrightarrow a \text{ and } b \text{ belong to the same coset of } K \qquad \textbf{(5-43)}$$

This is a consequence of Proposition 5-8. For if $f(a) = f(b)$, then $1 = f(a)f(b)^{-1} = f(a)f(b^{-1}) = f(ab^{-1})$. Thus, $ab^{-1} = k$, for some element k belonging to K. Then $a = kb \in Kb$, i.e., a and b belong to the same coset Kb. Reversing this argument shows conversely that if a and b belong to the same coset of K, then $f(a) = f(b)$.

The importance of Equation (5-43) is that it tells us there are as many images $f(a)$ as there are cosets of the kernel K of the homomorphism $f: G \to G'$. Since by definition, every element of G' is $f(a)$ for some a in G, it follows that

$$[G : K] = \text{the order of } G' \qquad \textbf{(5-44)}$$

5-9 Normal Subgroups and Factor Groups

For a subgroup H of a group G and an element x of G we may form the right coset Hx, as well as the left coset xH. There is no reason why these should be equal. In fact, the symmetric group S_3 contains the subgroup

$$H = \left\{\begin{pmatrix}123\\123\end{pmatrix}, \begin{pmatrix}123\\132\end{pmatrix}\right\}$$

If $x = \begin{pmatrix}123\\321\end{pmatrix}$, then

$$Hx = \left\{\begin{pmatrix}123\\321\end{pmatrix}, \begin{pmatrix}123\\312\end{pmatrix}\right\}$$

but

$$xH = \left\{\begin{pmatrix}123\\231\end{pmatrix}, \begin{pmatrix}123\\321\end{pmatrix}\right\}$$

and $Hx \neq xH$.

A subgroup H of a group G is said to be a *normal* subgroup if $Hx = xH$ for every x in G. We write $H \lhd G$ to indicate H is normal.

Evidently, if G is an Abelian group, then all its subgroups are normal.

The next theorem shows that normal subgroups arise in the study of homomorphism.

Theorem 5-9. The kernel K of a homomorphism $f\colon G \to G'$ is a normal subgroup of G.

Proof: We have already noted that K is a subgroup of G (see Proposition 5-9). To show $K \lhd G$, consider the cosets Kx and xK. If a is an arbitrary element of Kx, then $a = kx$ for some k in K. Certainly $a = x(x^{-1}kx)$; we claim that $x^{-1}kx$ belongs to K. Indeed $f(x^{-1}kx) = f(x^{-1})f(k)f(x) = f(x)^{-1}f(k)f(x) = f(x)^{-1}f(x)$ since $f(k) =$ the identity of G'. Thus, $f(x^{-1}kx) = 1$, as claimed.

We have shown that each element of Kx belongs to xK. A similar argument shown that any element of xK is also in Kx. Therefore, $Kx = xK$.

It is time to investigate more deeply the kernel K of a homomorphism $f\colon G \to G'$. We saw in Equation (5-43) that G' contains as many elements as there are cosets of K. That is, there is a correspondence

$$f(a) \leftrightarrow Ka$$

between the element $f(a)$ of G' and the coset Ka. If b is another element of G, then

$$f(b) \leftrightarrow Kb$$

and

$$f(ab) \leftrightarrow K(ab)$$

Since $f(ab) = f(a)f(b)$, the preceding correspondence suggests the further relation

$$K(ab) = (Ka)(Kb) \tag{5-45}$$

The key to the whole theory of homomorphism lies in the relation of Equation (5-45). We proceed as follows:

Let K be any normal subgroup of a group G. For cosets Ka and Kb, their *product* $(Ka)(Kb)$ is defined by the rule

$$(Ka)(Kb) = K(ab)$$

Since the cosets are equivalence classes and may have many different representations, it is necessary to show the preceding rule is well-defined. That is, we must show that if $Ka = Kc$ and $Kb = Kd$, then still $(Kc)(Kd) = K(ab)$. Since the product rule tells us $(Kc)(Kd) = K(cd)$, we must show, therefore, that $K(ab) = K(cd)$. In other words, given

$$Ka = Kc \quad \text{and} \quad Kb = Kd \tag{5-46}$$

we must show

$$K(ab) = K(cd) \tag{5-47}$$

The element ab belongs to the coset $K(ab)$. Also, since by Equation (5-46) $a \in Kc$ and $b \in Kd$, $a = kc$ and $b = k'd$ for some k, $k' \in K$. Then $ab = (kc)(k'd) = k(ck')d$. Now ck' belongs to the coset cK, and as $K \lhd G$, $cK = Kc$. Thus, $ck' = k''c$ for some $k'' \in k$. Therefore, $ab = k(ck')d = k(k''c)d = (kk'')(cd)$. The element kk'' belongs to K. Thus, $ab = (kk'')(cd)$ belongs to the coset $K(cd)$. Since the cosets $K(ab)$ and $K(cd)$ contain the element ab they are identical. That is, Equation (5-47) is valid.

With the product rule [Equation (5-45)] the set G/K of distinct right cosets of K forms a group, which is called the *factor group*. Since for a normal subgroup, its right and left cosets are identical, the elements of G/K are equally the left cosets of K. The identity of the factor group G/K is the coset K, and $(Ka)^{-1} = Ka^{-1}$. If G is a finite group, then the order of G/K is $[G : K]$.

We have seen earlier (Theorem 5-9) that the kernel of a homomorphism $f: G \to G'$ is a normal subgroup. Now we may prove the converse of this fact:

Theorem 5-10. If $K \lhd G$, then there is a homomorphism $f: G \to G/K$ such that $K =$ the kernel of f.

Proof: The elements of G/K are the cosets Kx and they are multiplied by the rule

$$(Ka)(Kb) = K(ab)$$

The homomorphism $f: G \to G/K$ is defined by

$$f(a) = Ka$$

Clearly, f maps G onto G/K. Moreover, for a,b in G, $f(ab) = K(ab) = (Ka)(Kb) = f(a)f(b)$. Therefore, f is a homomorphism.

Let $L =$ the kernel of f. If $a \in L$, then $f(a) = K = Ka$. Therefore, $a \in K$. So K contains L. On the other hand, L contains K, for if $k \in K$, then $Kk = K$; whence, $f(k) = K$. Altogether, $K = L$, as required.

5-10 The Fundamental Theorem of Homomorphism

In the last section we proved that the kernel of a homomorphism is a normal subgroup and conversely. The next result, which is known as the *Fundamental Theorem of Homomorphism* explains how completely the kernel of a homomorphism determines the homomorphic image.

Theorem 5-11. If K is the kernel of the homomorphism $f: G \to G'$, then G' is isomorphic to G/K.

Proof: The elements of G/K are the cosets Ka, while the elements of G' are the images $f(a)$, a in G.

We define a mapping $F: G/K \to G'$ by

$$F(Ka) = f(a) \quad \textbf{(5-48)}$$

That is, $F: Ka \to f(a)$. Manifestly, F is already onto. However, it has to be verified first of all that F is even a mapping. For the coset Ka may have a different representation $Ka = Kb$, and according to Equation (5-48), $F(Ka) = f(b)$ as well. Evidently, to show F is well defined, we must show that if

$$Ka = Kb \quad \textbf{(5-49)}$$

then

$$f(a) = f(b) \quad \textbf{(5-50)}$$

However, this is easy because Equation (5-49) implies $a = kb$ for some $k \in K$; whence, $f(a) = f(kb) = f(k)f(b) = f(b)$ (because k is in the kernel).

Now if Ka and Kb are elements of G/K, then $F(Ka \cdot Kb) = F(Kab) = f(ab) = f(a)f(b) = F(Ka)F(Kb)$. Thus,

$$F(Ka \cdot Kb) = F(Ka)F(Kb) \tag{5-51}$$

The mapping F, which is onto, is also a homomorphism. To prove G/K is isomorphic to G', it only remains to show F is one-to-one. Thus, suppose $F(Ka) = F(Kb)$. Then $f(a) = f(b)$ and $1 = f(a)f(b)^{-1} = f(a)f(b^{-1}) = f(ab^{-1})$. Therefore, $ab^{-1} \in K$ and $Ka = Kb$, which shows F is one-to-one.

The importance of Theorem 5-11 is that it permits one to construct *internally* all possible homomorphic images of a group G. Specifically, to describe all homomorphic images of G it is sufficient to

(i) find all normal subgroups K of G.
(ii) determine the factor groups G/K.

Let us carry out this program for the group I of integers. First, a word about notation: The operation of I is addition, so the cosets are written as $K + x$, rather than Kx. The "product" rule for the factor group I/K will read

$$(K + x) + (K + y) = K + (x + y)$$

Since I is Abelian, every subgroup of I is normal. Now if K is a nonzero subgroup of I, i.e., $K \neq \{0\}$, then K contains some nonzero integer u, and either u or $-u$ will be positive. Thus, K must contain at least one positive integer in the event $K \neq \{0\}$. Let m be the smallest positive integer in K. We claim

$$K = \{n \mid n \equiv 0 \pmod{m}\} \tag{5-52}$$

To see this, let n be any element of K. Then for some quotient q and remainder r.

$$n = mq + r$$

where $0 \leq r < m$. Since K is a subgroup containing m, K contains

$$mq = \underbrace{m + m + \ldots + m}_{q \text{ terms}}$$

Hence, K contains $-mq$ and $n - mq = r$. However, from the choice of m, K contains r only if $r = 0$, i.e., $n \equiv 0 \pmod{m}$.

Having found the normal subgroups of I, we consider next the structure of the factor groups I/K. If K is of the type Equation (5-52), then the coset decomposition of I relative to K is

$$I = K \cup K + 1 \cup \ldots \cup K + (m - 1)$$

Moreover, it is easy to see $K = [0]$, $K + 1 = [1], \ldots, K + (m - 1) = [m - 1]$. That is, the cosets are simply the equivalence classes of integers with respect to congruence modulo m.

The "product" rule in I/K is

$$(K + a) + (K + b) = K + (a + b)$$

or

$$[a] + [b] = [a + b]$$

Clearly, then, the factor group I/K is simply $I/(m)$.

Since by Theorem 5-11 the factor group $I/\{0\}$ is isomorphic to I, we have the following theorem:

Theorem 5-12. The homomorphic images of the group I of integers are (1) I itself and (2) $I/(m)$.

The structure of cyclic groups was determined in Theorems 5-4 and 5-5. Using the Fundamental Theorem of Homomorphism, these results may be reproved in a neat way.

Theorem 5-13. A cyclic group G is either isomorphic to the group I of integers or to one of the groups $I/(m)$.

Proof: For if a is a generator of G, then the mapping

$$f(n) = a^n \qquad n \in I$$

defines a homomorphism $f: I \to G$. If the kernel K of f is $\{0\}$, then f is an isomorphism. If $K \neq \{0\}$, then for some integer m,

$$K = \{n \mid n \equiv 0 \ (\text{mod } m)\}$$

and

$$G \cong I/K = I/(m)$$

5-11 Representations of Groups

Often questions about the structure of a group G are answered by first describing the group in a more familiar or concrete way. That is, one replaces the group G with some homomorphic image G', which is more tractable than G itself. Then the information obtained from examination of G' is transferred back to G. How successful one is in this transfer usually depends on the size or other properties of the kernel of the homomorphism $G \to G'$.

In this section we shall see two ways of representing a group as a permutation group. The first of these, which is called the *Cayley Representation Theorem*, was of historical importance for it showed that the notion of *abstract* group was not fundamentally different from the original ideal of a group as a set of permutations.

Theorem 5-14. *Cayley's Theorem.* Every group G is isomorphic to a subgroup of a suitable symmetric group.

To see the idea of the proof, it is convenient to look at an example. The cyclic group C of order 3 consists of the three elements

$$1, a, a^2$$

where $a^3 = 1$. Multiplying each of these elements by a simply permutes them, i.e., we obtain

$$a, a^2, 1$$

upon multiplication by a. If we denote this process of multiplication by R_a. then as a permutation of $1, a, a^2$

$$R_a = \begin{pmatrix} 1 & a & a^2 \\ a & a^2 & 1 \end{pmatrix}$$

Equally, multiplying the elements of $1, a, a^2$ by a^2 yields the permutation

$$R_{a^2} = \begin{pmatrix} 1 & a & a^2 \\ a^2 & 1 & a \end{pmatrix}$$

while

$$R_1 = \begin{pmatrix} 1 & a & a^2 \\ 1 & a & a^2 \end{pmatrix}$$

It is easy to see the three permutations R_1, R_a, R_{a^2} form a group C' which is cyclic and of order three. Therefore, C is isomorphic to the group C', which is a subgroup of S_3.

We now proceed to the proof of Theorem 5-14. For an element a of the group G let R_a be the mapping $R_a : G \to G$ defined by

$$R_a : x \to xa \qquad x \in G$$

This mapping turns out to be a permutation of the set G. Indeed, if $(x)R_a = (y)R_a$, then $xa = ya$; whence, $x = y$ by the cancellation law. Hence, R_a is a one-to-one map. Moreover, R_a is onto since if u is any element of G, then

$$u = (ua^{-1})R_a$$

Therefore, the set

$$G' = \{R_a \mid a \in G\}$$

consists entirely of permutations and G' is contained in S, the symmetric group of all permutations of the set G.

Observe that

$$R_aR_b = R_{ab} \quad \text{for all } a, b \text{ in } G \tag{5-53}$$

Indeed, if $x \in G$, then $(x)R_aR_b = ((x)R_a)R_b = (xa)R_b = (xa)b = x(ab) = (x)R_{ab}$. Thus, R_aR_b and R_{ab} are identical permutations.

It follows from Equation (5-53) that G' is a subgroup of S. Indeed, if R_a and R_b belong to G', then so does R_aR_b belong to G', by Equation (5-53). What remains to be shown is that $(R_a)^{-1} \in G'$ for all $a \in G$. However, R_1 clearly is the identity of S and by Equation (5-53) $R_aR_{a^{-1}} = R_1 = R_{a^{-1}}R_a$. Thus, $(R_a)^{-1} = R_{a^{-1}} \in G'$.

So far we have shown that the group G determines a subgroup G' of the symmetric group S. We show now that G is isomorphic to G'. This is done by setting up a mapping $f: G \to G'$ by the rule

$$f(a) = R_a$$

From the definition of G', the map $f: G \to G'$ is onto, and it is, therefore, a homomorphism because of Equation (5-53). Finally, f is one-to-one. For if $f(a) = f(b)$, then $R_a = R_b$; hence, $(x)R_a = (x)R_b$ for all x in G. Setting $x = 1$, we find $a = b$.

We have proved G and G' are isomorphic, and this completes the proof of the theorem.

Using the Cayley Representation Theorem we may derive the following result:

> Every group G of order $2p$ where p is an odd prime contains a normal subgroup H of index 2.

Certainly if G contains an element b of order p, then it generates a subgroup $H = \{1, b, \ldots, b^{p-1}\}$ which is of order p and index 2, hence, is normal in G (see Exercise 5-9, Problem 1). Equally, if G contains an element c of order $2p$, then $b = c^2$ will be of order p, and G will contain again a normal subgroup H of index 2.

If G contains an element u of order two, then we use the Cayley Representation

$$a \rightarrow R_a$$

of G as a permutation group. Since u has order two, R_u, in cycle form, is a product of p transpositions. Thus, the group $G' = \{R_a \mid a \in G\}$ contains an odd transposition. By arguing as in the proof of Proposition 5-5 it may be shown that the set H of all even permutations of G' forms a subgroup of index 2.

The value of the Cayley Representation Theorem is that it makes available to arbitrary groups facts about permutations, cycles, etc. The disadvantage of the theorem is that when a group G of order n is represented as a subgroup of S_n, G, so to speak, becomes lost in S_n because S_n has such large order compared to G. For example, representing a group G of order 8 as a group G' of permutations, G' is contained in S_8 which has order 40,320.

An instructive example of the deficiency of the representation provided by Cayley's Theorem is the group G of isometries of the square. It was shown earlier (in Section 5-4) that G is of order 8. The Cayley Theorem tells us the elements of G may be regarded as permutations of 8 symbols. However, the representation is unnatural from the geometric point of view since G is thought of as a permutation group on the *four* vertices of the square.

The problem of finding representations of groups other than those given in Theorem 5-14 in a major topic is the theory of groups and is much too extensive to be given here. However, we shall point out that as a generalization of Theorem 5-14, a group may be represented as a permutation on the cosets of any of its subgroups.

If G is a finite group and H is a subgroup of G, then decomposing G into distinct cosets of H

$$G = H \cup Hx_2 \cup \ldots \cup Hx_m$$

For an element x in G the cosets

$$Hx, Hx_2x, \ldots, Hx_mx$$

will be the original cosets

$$H, Hx_2, \ldots, Hx_m$$

in perhaps some different order. Thus, each x of G determines a permutation P_x of the set $X = \{H, Hx_2, \ldots, Hx_m\}$.

Explicitly,

$$P_x = \begin{pmatrix} H & Hx_2 & \ldots & Hx_m \\ Hx & Hx_2x & & Hx_mx \end{pmatrix}$$

For permutations P_x, P_y and a typical coset Hx_i, $(Hx_i)P_xP_y = ((Hx_i)x)y = Hx_i(xy) = (Hx_i)P_{xy}$.

Thus,

$$P_xP_y = P_{xy}$$

for all x, y in G, and the mapping f

$$f: x \to P_x$$

defines a homomorphism $f: G \to G'$, where

$$G' = \{Hx \mid x \in G\}$$

In this way, we have represented G as a subgroup of S_m, where $m = [G : H]$. The homomorphism $f: G \to G'$ need not be an isomorphism. Its kernel K consists of those $x \in G$ for which P_x is the identity, and this occurs when for each coset Ha, $Hax = Ha$. Therefore, $x \in K \Leftrightarrow axa^{-1} \in H$ for all $a \in G$. In particular, for $a = 1$, we find $x \in H$. Therefore, the kernel K of the homomorphism f is a normal subgroup of G, which is contained in H, and $K = \{x \in H \mid axa^{-1} \in H \text{ for all } a \in G\}$.

Comparing this representation with that of the Cayley Theorem, we see that a group G order n is *isomorphic* to a subgroup of S_n, while if G has a subgroup H of order t, then G is *homomorphic* to a subgroup of $S_{n/t}$. In the latter case, we have representation of G in a smaller symmetric group but may have to sacrifice the isomorphism in achieving this representation.

Example 5-6. Let G be the group of order 8 with multiplication Table 5-8.

G has a subgroup $H = \{1,x_5\}$ of order 2. There are 4 right cosets of H, and denoting these by the letters A, B, C, D, we have

$$\begin{aligned} A &= H = \{1,x_5\} \\ B &= Hx_2 = \{x_2,x_8\} \\ C &= Hx_3 = \{x_3,x_7\} \\ D &= Hx_4 = \{x_4,x_6\} \end{aligned}$$

It is easy to see that H is not a normal subgroup of G; hence, when representing G as a permutation group on the cosets of H, we will, in

Table 5-8

	1	x_2	x_3	x_4	x_5	x_6	x_7	x_8
1	1	x_2	x_3	x_4	x_5	x_6	x_7	x_8
x_2	x_2	x_3	x_4	1	x_6	x_7	x_8	x_5
x_3	x_3	x_4	x_2	1	x_7	x_8	x_5	x_6
x_4	x_4	1	x_2	x_3	x_8	x_5	x_6	x_7
x_5	x_5	x_8	x_7	x_6	1	x_4	x_3	x_2
x_6	x_6	x_5	x_8	x_7	x_2	1	x_4	x_3
x_7	x_7	x_6	x_5	x_8	x_3	x_2	1	x_4
x_8	x_8	x_7	x_6	x_5	x_4	x_3	x_2	x_1

fact, have an isomorphic from G onto a subgroup of S_4. Calculating P_{x_2} and P_{x_5} we find

$$P_{x_2} = \begin{pmatrix} A & B & C & D \\ B & C & D & A \end{pmatrix}$$

and

$$P_{x_5} = \begin{pmatrix} A & B & C & D \\ A & D & C & B \end{pmatrix}$$

Moreover,

$$P_{x_5}P_{x_2} = (P_{x_2})^3 P_{x_5}$$

From this last relation and the fact that P_{x_2} and P_{x_5} are of orders 4 and 2 respectively, it is not difficult to see that G is isomorphic to the isometry group of the square of Figure 5-9. The permutation P_{x_2} corresponds to a counter-clockwise rotation of 90° about the center 0 of the square, while P_{x_5} corresponds to the reflection of the square about the axis $A'C'$.

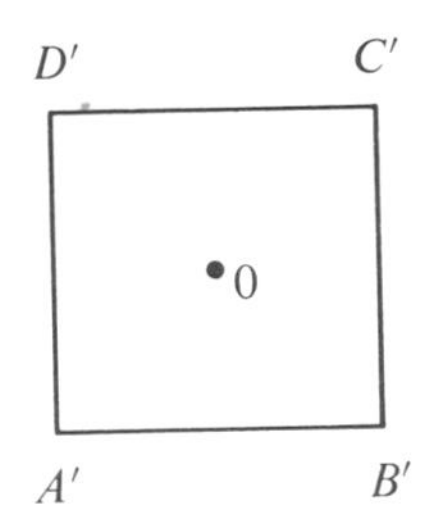

Figure 5-9

5-12 Conjugates and the Class Equation

An isomorphism of a group G onto *itself* is called an automorphism. It is easy to construct automorphisms of groups.

Remark 5-5. For a group G and an element x of G, the mapping

$$\phi : a \to x^{-1}ax$$

is an automorphism of G.

Indeed, $\phi(ab) = x^{-1}abx = (x^{-1}ax)(x^{-1}bx) = \phi(a)\phi(b)$, and for any $u \in G$, $u = \phi(xux^{-1})$. Hence, $\phi : G \to G$ is onto and preserves products, i.e., ϕ is a homomorphism. If $\phi(a) = \phi(b)$, then $x^{-1}ax = x^{-1}bx$, and after canceling, $a = b$. Thus, ϕ is one-to-one, which completes the verification of Remark 5-5.

If H is a subgroup of G, then its image $\phi(H) = x^{-1}Hx$ under the automorphism (12.1) is also a subgroup (see Exercise 5-9, Problem 2). The subgroup $x^{-1}Hx$ is called to be a *conjugate* of H. More generally, the element $x^{-1}ax$ is said to be a *conjugate* of a. A normal subgroup is one which is equal to all its conjugates (see Exercise 5-9, Problem 2).

Proposition 5-10. For a group G the relation $\backsim$ defined by

$$a \backsim b \quad \text{if} \quad a = x^{-1}bx \quad \text{for some } x \text{ in } G$$

is an equivalence relation.

Proof: The relation $\backsim$ is reflexive since $a = a^{-1}aa$ for all $a \in G$, and it is symmetric since if $a = x^{-1}bx$, then $b = xax^{-1} = (x^{-1})^{-1}ax^{-1}$. Moreover, if $a \backsim b$ and $b \backsim c$, then $a = x^{-1}bx$ and

$b = y^{-1}cy$ for some x,y in G; hence, $a = x^{-1}y^{-1}cyx = (yx)^{-1}c(yx)$, which shows $\backsim$ is also transtive.

The equivalence relation $\backsim$ partitions the group G into distinct equivalence classes; the equivalence class containing a is called the *conjugate class* of a and is denoted by C_a.

In order to count the number of elements in the conjugate class C_a, we form first the *centralizer* $C_G(a)$. This is defined to be the set all elements of G which commute with a. That is,

$$C_G(a) = \{x \in G \mid xa = ax\}$$

Clearly

$$C_G(a) = \{x \in G \mid x^{-1}ax = a\}$$

as well.

It is easy to see that $C_G(a)$ is a subgroup of G, and from this fact we count the number of elements in the conjugate class C_a.

If two conjugates of a are equal, for instance $x^{-1}ax = y^{-1}ay$, then $(yx^{-1})a - a(yx^{-1})$ and $yx^{-1} \in C_G(a)$. Therefore, y and x belong to the same right coset $C_G(a)x$ of the subgroup $C_G(a)$. Conversely, if x and y belong to the same right coset of $C_G(a)$, then $xy^{-1} \in C_G(a)$ and $x^{-1}ax = y^{-1}ay$. So, there are as many conjugates of a as there are cosets of $C_G(a)$. Therefore, for a finite group G *the conjugate class* C_a contains $c_a = [G : C_G(a)]$ elements. Consequently, c_a is a divisor of the order of G.

Now, if G is a group of order n, then it is partitioned into conjugate classes $C_a, C_b, \ldots, C_y$ and counting the elements of G, we have

$$n = c_a + c_b + \ldots + c_g \qquad \textbf{(5-54)}$$

where the integers $c_a, c_b, \ldots, c_g$ are divisors of n.

Equation (5-54) is called the *class equation* of G.

The center Z of a group G is defined to consist of those elements of G which commute with every element of G, that is,

$$Z = \{z \in G \mid zx = xz \quad \text{for all } x \in G\}$$

Clearly, Z is a normal subgroup of Z. Moreover, in Equation (5-54), $c_a = 1$ if and only if $a \in Z$.

5-13 The Theorems of Cauchy and Sylow

Lagrange's Theorem states that if G is a finite group of order n and H is a subgroup of order t, then t divides n. The converse of this theorem is generally false. For example, the alternating group A_4 has order 12, but there is no subgroup of A_4 which has order 6.

In this section we prove some results which show that for certain divisors d of the order of a finite group G there does exist subgroups of order d.

Theorem 5-15. If G is an Abelian group of order n, then for each prime p dividing n, G contains a subgroup H of order p.

Proof: The argument we are about to give is typical of the way in which factor groups are used.

The proof is by induction on the integer m where $n = pm$ is the order of G. If $m = 1$ we may choose $H = G$.

Assuming the result is true for all groups of order pm' where $m' < m$, let G be a group of order pm and choose an element $a \neq 1$ of G. If the order t of a is divisible by p, then $t = pk$ for some integer $k \geq 1$ and the element $b = a^k$ will have order p and generate a subgroup

$$H = \{1, b, \ldots, b^{p-1}\}$$

of order p.

Consequently, let us assume the order t of a is not divisible by p. Then,

$$K = \{1, a, \ldots, a^{t-1}\}$$

is a subgroup of G and $K \lhd G$ (because G is Abelian). The factor group G/K has order $p(m/t)$, and as $m/t < m$, the induction hypothesis guarantees the existence of an element of order p in the factor group. Such an element is a coset Kb where $Kb \neq K$ and $(Kb)^p = K$; hence,

$$b \notin K \qquad b^p \in K$$

From the last of these relations and the fact that K has order t, $b^{pt} = 1$ (see Corollary 5-2). Put $c = b^t$. Then, $c^p = b^{pt} = 1$. Moreover, $c \neq 1$. For if $c = 1$, then $(Kb)^t = 1$, and as $Kb \neq 1$, the order p of the coset Kb would have to divide t, which is contrary to assumption.

The element c we have constructed is, therefore, of order p and the subgroup

$$H = \{1, c, \ldots, c^{p-1}\}$$

it generates has order p.

The next theorem, which is an extension of the previous result, is due to Cauchy.

Theorem 5-16. If G is a group of order $n = pm$, where p is a prime, then G contains a subgroup H of order p.

Proof: For this result we require the class Equation (5-54) for G. Recall that the class $C_a = \{a\} \Leftrightarrow a$ is contained in the center Z. In view of this, we may write the class equation in the form

$$n = [Z : 1] + \sum_{a \notin Z} c_a \tag{5-55}$$

where

$$c_a = [G : C_G(a)] \tag{5-56}$$

Now, if p divides each of the integers c_a, $a \notin Z$, then p divides $[Z : 1]$. However, Z is an Abelian group, and, therefore, contains a subgroup H of order p because of the previous theorem.

If, on the other hand, some c_a, $a \notin Z$, is not divisible by p, then p divides $[C_G(a) : 1]$ because of Equation (5-56). As $a \notin Z$, the subgroup $C_G(a) \neq G$; hence, it has order pm' where $m' < m$. By induction, $C_G(a)$ contains a subgroup of order p.

The following theorem, which sharpens Cauchy's result is one of the most important in the theory of finite groups.

Theorem 5-17. *Sylow's Theorem.* If G is a group of order $p^\alpha m$ where p is a prime and $(p,m) = 1$, then G contains a subgroup P of order p^α.

Proof: We use induction on the order of G, the result, being trivial if $[G : 1] = p$ or $\alpha = 0$. Assume the theorem is true for all groups of order $< p^\alpha m$ and let G be of order $p^\alpha m$.

As before, we consider the class equation

$$p^\alpha m = [Z : 1] + \sum_{a \notin Z} c_a \qquad c_a = [G : C_G(a)]$$

where $Z =$ the center of Z. If for some $a \notin Z$, $c_a \not\equiv 0 \pmod{p}$, then $C_G(a)$ is a subgroup of order $p^\alpha m'$, where $m' < m$, and by the induction hypothesis it contains a subgroup P of order p^α.

Therefore, we assume $c_a \equiv 0 \pmod{p}$ for all $a \notin Z$. It follows from the class equation that the center Z is of order divisible by p. As Z is Abelian, it contains a subgroup K of order p, and K is clearly a normal subgroup of G. Then the factor group G/K, which is of order $mp^{\alpha-1}$, contains a subgroup P' of order $p^{\alpha-1}$.

Now we know there is a homomorphism $f: G \to G/K$ whose kernel is K (see Theorem 5-10). Let $P = \{x \in G \mid f(x) \in P'\}$. Evidently, P is a subgroup of G. Trivially, P contains K. Decomposing P into distinct cosets relative to the subgroup K, we have

$$P = K \cup Kx_2 \cup \ldots \cup Kx_s$$

whence, $P' = \{f(1), f(x_2), \ldots, f(x_s)\}$, where $s = p^{\alpha-1}$. Therefore, $[P : K] = p^{\alpha-1}$ and $[P : 1] = [P : K][K : 1] = p^{\alpha-1}p = p^{\alpha}$, as required.

5-14 Translation Groups

By a *theory*, one generally means an explanation of a phenomenon from some point of view. The theory of groups we have given in this chapter explains groups in terms of subgroups, cosets, and homomorphisms.

In this section we give the beginnings of a theory of geometry based on groups. The advantage of such a theory should be obvious — it permits the application of the powerful theory of groups to geometric problems.

An *endomorphism* of an Abelian group A is mapping $\phi : A \to A$ such that

$$\phi(ab) = \phi(a)\phi(b) \quad \text{for all } a,b \in A$$

An endomorphism ϕ of A would be a homomorphism $\phi : A \to A$ except for the fact that ϕ need not be an onto mapping. If ϕ_1 and ϕ_2 are endomorphism of A, then as mappings of A, their *product* $\phi_1\phi_2$ is given by

$$\phi_1\phi_2(a) = \phi_1(\phi_2(a)) \qquad a \in A \tag{5-57}$$

The product $\phi_1\phi_2$ is again an endomorphism of A, since for a,b in A,

$$\begin{aligned}\phi_1\phi_2(ab) &= \phi_1(\phi_2(ab)) = \phi_1(\phi_2(a)\phi_2(b)) = \phi_1(\phi_2(a))\phi_1(\phi_2(b)) \\ &= \phi_1\phi_2(a)\phi_1\phi_2(b)\end{aligned}$$

The *sum* $\phi_1 + \phi_2$ of endomorphisms ϕ_1 and ϕ_2 of an Abelian group A is defined by

$$(\phi_1 + \phi_2)(a) = \phi_1(a)\phi_2(a) \qquad a \in A \tag{5-58}$$

The mapping $\phi_1 + \phi_2$ is also an endomorphism of A. For if a and b belong to A, then

$$\begin{aligned}(\phi_1 + \phi_2)(ab) &= \phi_1(ab)\phi_2(ab) \\ &= \phi_1(a)\phi_1(b)\phi_2(a)\phi_2(b) \\ &= \phi_1(a)\phi_2(a)\phi_1(b)\phi_2(b) \quad \text{(because } A \text{ is abelian)} \\ &= (\phi_1 + \phi_2)(a)(\phi_1 + \phi_2)(b)\end{aligned}$$

Now, let F be a field and Π be the affine geometry which is coordinatized by F. A *translation* of Π is a mapping T of the points of Π such that

$$T(x,y) = (x + a, y + b) \qquad \textbf{(5-59)}$$

for all points (x,y) of Π, where a and b are fixed quantities of F. Since the translation T depends upon a and b, we write $T = T_{a,b}$ to denote this fact. Clearly, $T_{a,b}$ is a permutation of the points of Π. Moreover,

$$T_{a,b}T_{c,d} = T_{a+c,b+d} \qquad \textbf{(5-60)}$$

and

$$(T_{a,b})^{-1} = T_{-a,-b} \qquad \textbf{(5-61)}$$

for all a,b,c and d in F.

It follows from Equations (5-60) and (5-61) that the set A of translations of Π is a subgroup of the group of all permutations of the points of Π. The group A is called the *translation group* of Π. Clearly, A is an Abelian group.

It will turn out that the geometry Π can be completely described by its translation group.

For each $k \in F$ we construct an endomorphism ϕ_k of the group A. This endomorphism is defined by

$$\phi_k : T_{a,b} \rightarrow T_{ak,bk} \qquad \textbf{(5-62)}$$

Let us verify ϕ_k is an endomorphism. For translations $T_{a,b}$ and $T_{c,d}$, $T_{a,b}T_{c,d} = T_{a+c,b+d}$; hence,

$$\phi_k(T_{a,b}T_{c,d}) = T_{(a+c)k,(b+d)k}$$

by Equation (5-62). On the other hand,

$$\begin{aligned}\phi_k(T_{a,b})\phi_k(T_{c,d}) &= T_{ak,bk}T_{ck,dk} \\ &= T_{ak+ck,bk+dk} \quad \text{(by Equation (5-60)}\end{aligned}$$

therefore, $\phi_k(T_{a,b}T_{c,d}) = \phi_k(T_{a,b})\phi_k(T_{c,d})$, and ϕ_k is an endomorphism of A.

Let

$$F^* = \{\phi_k \mid k \in F\} \tag{5-63}$$

As the set F^* of endomorphisms is constructed from the field F, let us indicate this by putting

$$\phi_k = \Phi(k) \tag{5-64}$$

Evidently $\Phi : k \to \phi_k$ is a mapping. Clearly, the mapping $\Phi : F \to F^*$ is onto.

Now if $k,k' \in F$, then the endomorphisms ϕ_k and $\phi_{k'}$ are

$$\phi_k(T_{a,b}) = T_{ak,bk}$$

and

$$\phi_{k'}(T_{a,b}) = T_{ak',bk'}$$

For the product $\phi_k\phi_{k'}$ of these endomorphisms,

$$\phi_k\phi_{k'}(T_{a,b}) = \phi_k(\phi_{k'}(T_{a,b})) = \phi_k(T_{ak',bk'}) = T_{ak'k,bk'k} = \phi_{kk'}(T_{a,b})$$

Consequently, $\phi_k\phi_{k'}$ and $\phi_{kk'}$ are the same endomorphism of A. Hence, by the definition [Equation (5-64)] of Φ, we have

$$\Phi(kk') = \Phi(k)\Phi(k') \quad \text{for all } k,k' \in F \tag{5-65}$$

Similarly, for the sum $\phi_k + \phi_{k'}$ of the endomorphisms ϕ_k and $\phi_{k'}$,

$$\begin{aligned}(\phi_k + \phi_{k'})(T_{a,b}) &= \phi_k(T_{a,b})\phi_{k'}(T_{a,b}) = T_{ak,bk}T_{ak',bk'} = T_{ak+ak',bk+bk'}\\ &= T_{a(k+k'),b(k+k')} = \phi_{k+k'}(T_{a,b}).\end{aligned}$$

Hence,

$$\Phi(k + k') = \Phi(k) + \Phi(k') \quad \text{for all } k,k' \in F \tag{5-66}$$

Finally, the mapping $\Phi : F \to F^*$ is one-to-one. For if $\Phi(k) = \Phi(k')$, then the endomorphisms ϕ_k and $\phi_{k'}$ are equal, which implies $T_{ak,bk} = T_{ak',bk'}$ for all $a, b \in F$. In particular, for $a = b = 1$, $T_{k,k} = T_{k',k'}$; whence, $k = k'$, i.e., Φ is one-to-one.

Combining the information given we have the following result:

Remark 5-6. The mapping $\Phi : F \to F^*$ is an isomorphism. In particular, F^* is a field.

What is the significance of the preceding statement? The affine geometry Π is determined by the field F which coordinatizes Π. However, the statement shows that F is determined, up to isomorphism, by the translation group A of the geometry Π. Consequently, the geometry Π is determined, up to isomorphism, by its translation group A.*

The translation group may be defined without referring to coordinates, i.e., it may be described completely geometrically. For a translation T, other than the identity, is characterized by the following two properties:

Property 5-8. For each pair of distinct points P and Q of Π, the line joining P and Q is parallel to the line joining $T(P)$ and $T(Q)$.

Property 5-9. $T(P) \neq P$ for all points P of Π.

The field F^* of endomorphisms (Equation (5-63)) may also be described purely geometrically. To see this, note first that the translation $T_{a,b}$ can be assigned a definite direction if either $a \neq 0$ or $b \neq 0$. The mapping $T_{a,b}$ translates the original coordinate system to one whose center is (a,b) (see Figure 5-10). The direction of the translation $T_{a,b}$ is specified by the vector $[a,b]$. In fact, the direction of a translation $T_{a,b}$ may be defined without reference to the field F at all by agreeing with the following definition:

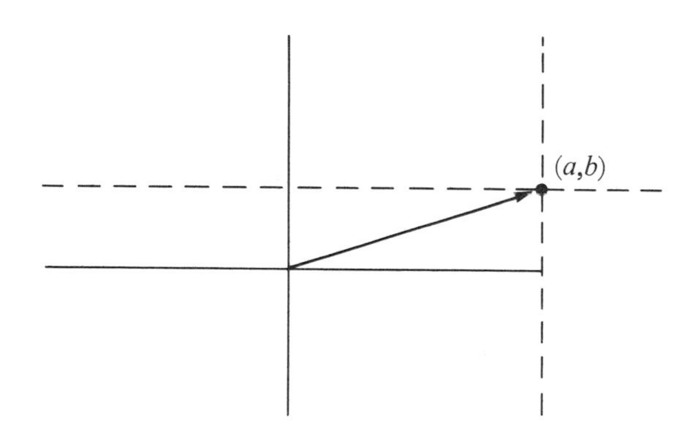

Figure 5-10

Definition 5-1. The *direction* of a translation $T \neq 1$ is the line joining P and $T(P)$, where P is some point of Π.

*For the definition of isomorphism of affine geometries, see Exercise 2-7, Problem 6.

Property 5-8 guarantees that the Definition 5-1 is independent of the choice of the point P. We do not assign any direction to the identity mapping. Now what are the endomorphisms of Equation (5-62)? These are precisely the endomorphisms of the translation group which leave the directions of the translations unchanged.

What we have proved is the following theorem:

Theorem 5-18. If Π is the affine geometry coordinatized by a field F, then F is isomorphic to the field of all direction-preserving endomorphisms of the translation group of Π.

In Section 2-5 we mentioned that an abstract affine geometry can be coordinatized by an algebraic system. Now we can be more explicit about this and explain how the coordinatizing system is constructed.

Let Π be an abstract affine geometry. Hence, Π is a system of objects called points and sets of these points called lines such that the following are true:

Axiom 5-6. Given any pair of distinct points A and B of Π, there exists a unique line of Π containing A and B.

Axiom 5-7. For any line l of Π and point P not on l there exists exactly one line l' such that l' passes through P and does not intersect l.

Axiom 5-8. Each pair of distinct lines of Π either do not intersect or they intersect in exactly one point.

Assume further that Π is *nondegenerate* in the sense of the following:

Axiom 5-9. Π contains three noncollinear points.

A *translation* of Π is a permutation T of the points of Π such that either $T = 1$ or T satisfies Properties 5-8 and 5-9. The *translation group* of Π is the group A of all translations.

We have to assume further that A contains enough translations, i.e., the following:

Axiom 5-10. For each pair of points P, Q of Π there exists a (unique) translation T such that $T(P) = Q$.

In Section 5-4 we indicated that Axiom 5-10 is equivalent to Axiom 5-11.

Axiom 5-11. Desargue's Theorem is valid for the geometry Π.

Once the affine geometry satisfies the conditions Axioms 5-6 to 5-11, it can be proved that the set F^* of direction-preserving endomorphisms of the translation group of A is an algebraic system satisfying all the axioms for a field, except for the commutative law of multiplication. The geometry Π is coordinatized by this system F^*. For the details of the proof of these facts, we refer the reader to *Lectures on Modern Geometry;* Segre, B.; Edivioni Cremonese; Rome, Italy; 1961.

Given an Abelian group A, does there exist an affine geometry whose translation group is isomorphic to A? The following example shows the kind of analysis necessary to answer such a question.

Let A be the noncyclic group of order 4. Its Cayley table is Table 5-9.

Table 5-9

	1	a	b	c
1	1	a	b	c
a	a	1	c	b
b	b	c	1	a
c	c	b	a	1

If A is to be the translation group of a geometry Π, then there should be as many points in Π as there are elements of A [see Equation (5-59)], and the elements of A should be permutations of the points of Π.

Now, the Cayley Representation Theorem tells us A is isomorphic to a subgroup of S_4. Calculating the representation $x \longrightarrow R_x$, $x = 1, a, b,$ and c, we find the representation

$$1 \longrightarrow R_1 = \begin{pmatrix} 1 & a & b & c \\ 1 & a & b & c \end{pmatrix}$$

$$a \longrightarrow R_a = \begin{pmatrix} 1 & a & b & c \\ a & 1 & c & b \end{pmatrix}$$

$$b \longrightarrow R_b = \begin{pmatrix} 1 & a & b & c \\ b & c & 1 & a \end{pmatrix}$$

$$c \longrightarrow R_c = \begin{pmatrix} 1 & a & b & c \\ c & b & a & 1 \end{pmatrix}$$

In these permutations replace the symbols 1, a, b, and c with the letters P, Q, R, and S respectively. Then A is isomorphic to the group

$$A' = \left\{1, \begin{pmatrix} P & Q & R & S \\ Q & P & S & R \end{pmatrix}, \begin{pmatrix} P & Q & R & S \\ R & S & P & Q \end{pmatrix}, \begin{pmatrix} P & Q & R & S \\ S & R & Q & P \end{pmatrix}\right\} \tag{5-67}$$

We take the points of Π to be the letters P, Q, R, and S. Now, if X and Y are any two distinct points of Π, then we should have a permutation T of A' which maps X onto Y (see Axiom 5-10) and the line l joining X and Y should be

$$l = \{X, T(X)\}$$

Taking $X = P, Q, R$, and S, and letting T be the four permutations of A', we find that the lines of Π are

$$\{P,Q\}, \{P,R\}, \{P,S\}, \{Q,S\}, \{Q,R\}, \{R,S\}$$

Exercises

Exercise 5-1

1. Write down the multiplication table for the multiplicative group of all nonzero elements of the field $I/(7)$.

2. Find the multiplication tables for the groups of order 4. Show there are exactly two nonisomorphic groups of order 4.

3. Show that for a group G

$$(ab)^{-1} = b^{-1}a^{-1}$$

   for all $a,b \in G$.

4. Show that a group G is Abelian if and only if

$$(ab)^{-1} = a^{-1}b^{-1}$$

   for all $a,b \in G$.

5. Show that if $x^2 = 1$ for all $x \in G$, then G is an Abelian group.

6. Let G denote the set of all real numbers $\neq 1$. Define an operation $a \circ b$ in G in terms of the usual addition and multiplication by the rule

$$a \circ b = a + b - ab$$

   Verify that G becomes a group with respect to the preceding rule of combination.

7. Let G and G' be groups and $f: G \rightarrow G'$ be a one-to-one onto mapping such that

$$f(ab) = f(b)\,f(a)$$

for all $a,b \in G$. Show that G and G' are isomorphic groups. (Hint: Consider the mapping $a \to f(a^{-1})$.)

8. Let a and b be elements of a group G such that $a^4 = b^2 = 1$ and $ba = a^3b$. Find an integer n such that $ba^3 = a^nb$.

Exercise 5-2

1. Write each of the following permutations in their disjoint cycle form:

$$\begin{pmatrix}12345\\23154\end{pmatrix} \quad \begin{pmatrix}12345\\13452\end{pmatrix} \quad \begin{pmatrix}123456\\456123\end{pmatrix} \quad \begin{pmatrix}A\ B\ C\ D\ E\ F\ G\\A\ C\ B\ D\ E\ G\ F\end{pmatrix}$$

2. Show that two cycles u and v commute in the sense that $uv = vu$ if and only if they have no element in common.

3. Determine all permutations σ of S_4 such that $\sigma\tau = \tau\sigma$, where

$$\tau = \begin{pmatrix}1\ 2\ 3\ 4\\2\ 3\ 4\ 1\end{pmatrix}$$

4. Find an even permuation σ of S_6 for which

$$(1)\ \sigma = 3 \qquad (2)\ \sigma = 6 \qquad (3)\ \sigma = 4$$

5. Find an odd permutation σ of S_6 for which

$$(1)\ \sigma = 3 \qquad (2)\ \sigma = 6 \qquad (3)\ \sigma = 4$$

Exercise 5-3

1. Show that if H and K are subgroups of a group G, then so is their intersection $H \cap K$ a subgroup of G.

2. Suppose H and K are subgroups of G such that $[H:1] = m$ and $[K:1] = n$, where m and n are relatively prime integers. Determine the order of $H \cap K$.

3. Let G be the multiplicative group of all complex numbers z for which

$$z^{12} = 1$$

   (a) Show that the set H of all complex numbers w such that $w^3 = 1$ is a subgroup of G.
   (b) Decompose the group G into right cosets relative to the subgroup H.
   (c) Determine all subgroups of G.

4. Find subgroups H and K of S_3, which have orders 2 and 3 respectively. Decompose S_3 into right cosets relative to H.

5. Let G be the multiplicative group of all nonzero elements of the field $I/(13)$. Find the subgroups of G.

6. Let G be a group and a and b be elements of G such that $a^4 = b^2 = 1$, and $ba = a^3b$. Verify the following set H is a subgroup of G:

$$H = \{1,a,a^2,a^3,b,ab,a^2b,a^3b\}$$

7. Let $f: G \to G'$ be an isomorphism of groups G and G'; let H be a subgroup of G. Show that the set
$$f(H) = \{f(h) \mid h \in H\}$$
is a subgroup of G'.

8. Let G be a group and $a \in G$.
   (a) Show that the set
$$H = \{x \in G \mid ax = xa\}$$
   is a subgroup of G.
   (b) For the group $G = S_4$ and $a = \begin{pmatrix} 1 & 2 & 3 & 4 \\ 2 & 1 & 4 & 3 \end{pmatrix}$ calculate H and decompose G into right cosets relative to H.

9. Show that in general Theorem 5-1 is false for an *infinite* subset H.

Exercise 5-4

1. Describe the isometry group of a circle.

2. Calculate the isometry group of a rectangle which is not a square.

3. Can the isometry group of a triangle be isomorphic to the isometry group of a quadrilateral?

4. The *dihedral* group D_n is defined to be the isometry group of a regular n-gon. Show that D_n has order $2n$.

5. Calculate the isometry group of the *Star of David*.

6. Let σ be an isometry of the plane having the property that for each pair of distince points A and B, the lines AB and $(A)\sigma(B)\sigma$ are parallel. Show that if $(P)\sigma = P$ for some point P, then $\sigma = 1$.

Exercise 5-5

1. Determine anew the possible groups of order 4 by considering the possible orders of the elements of such a group.

2. Let a and b be elements of orders m and n respectively of an Abelian group. Show that if m and n are relatively prime, then ab has order mn.

3. Let G be an Abelian group which has elements of orders 6, 11, and 35. What is the smallest possible order G could have?

4. Let G be a group and $a \in G$ have order mn, where m and n are relatively prime integers. Show there exist $b, c \in G$ of orders m and n respectively such that $a = bc$. (Hint: Choose b and c to be suitable powers of a.)

5. The following is a partial multiplication table for a group of order 6. Determine the missing entries by considering the order of the element a.

	1	a	b	c	d	e
1	1	a	b	c	d	e
a	a	b				
b	b				c	d
c	c					
d	d		e			
e	e			b		

6. Prove that a cycle $(a_1a_2 \ldots a_k) \epsilon S_n$ has order k. Then show that when an element of S_n is written in its disjoint cycle form, its order is the least common multiple of the lengths of its cycles.

7. Prove that a group of order 6 must contain an element of order 3.

8. Prove that cyclic groups are Abelian.

9. Prove that a cycle group of order n has exactly one subgroup of order m for each divisor m of n.

10. Show that subgroups of cyclic groups are always cyclic.

11. Let G be a cyclic group of order n. Show that G contains $\phi(n)$ elements of order n.

Exercise 5-6

1. Of the groups G_n, where $n \leq 12$, determine which are cyclic.

2. If A and B are groups, their *direct product* $A \times B$ is defined to consist of the ordered pairs

$$\{(a,b) \mid a \epsilon A, b \epsilon B\}$$

Show that $A \times B$ becomes a group under the multiplication rule

$$(a,b)(c,d) = (ac,bd)$$

3. Let $(m,n) = 1$. Show that G_{mn} is isomorphic to $G_m \times G_n$. (Hint: Use the Chinese Remainder Theorem.)

Exercise 5-7

1. An *isometry* of space is defined in much the same way as an isometry of the plane. That is, it is just a permutation of the points of space which preserves distances. If $\mathfrak{C}$ is a figure in space, then its *isometry group* G consists of those isometries which map $\mathfrak{C}$ onto $\mathfrak{C}$. Show that the isometry group G of a cube $\mathfrak{C}$ has order 48 by considering the subgroup of G which fixes one of the faces of $\mathfrak{C}$.

2. Show that if G is a non-Abelian group of order 8, then G contains an element of order 4. Then show there are exactly two nonisomorphic, non-Abelian groups of order 8.

3. Calculate the order of the collineation group of the affine geometry which is coordinatized by the field $I/(3)$.

4. Determine all groups of order 10.

Exercise 5-8

1. Let G be a group of order 12 and $f: G \rightarrow G'$ be a homomorphism, where G' is a non-Abelian group. Show that if the kernel of f contains an element of order 3, then G' is isomorphic to S_3.

2. Let $f: G \rightarrow G'$ be a homomorphism of groups and H be a subgroup of G. Show that

$$f(H) = \{ f(h) \mid h \in H\}$$

is a subgroup of G'.

3. For an element $\sigma \in S_n$, put $f(\sigma) = +1$ if σ is even and $f(\sigma) = -1$ if σ is odd. Show that

$$f: S_n \rightarrow G'$$

is a homomorphism, where $G' = \{+1, -1\}$. Use this to show that the alternating group A_n has order $n!$.

4. Show that if G, G', and G'' are groups for which there are homomorphisms

$$f: G \rightarrow G'$$
$$g: G' \rightarrow G''$$

then there exists a homomorphism $h: G \rightarrow G''$.

5. Show that if $f: G \rightarrow G'$ is a homomorphism and G is cyclic, then G' is cyclic.

Exercise 5-9

1. Show that if H is a subgroup of G such that $[G : H] = 2$, then $H \lhd G$.

2. For a subgroup H of a group G and $x \in G$ let

$$x^{-1}Hx = \{x^{-1}hx \mid h \in H\}$$

   Show that H is a normal subgroup of G if and only if $x^{-1}Hx = H$ for all $x \in G$.

3. Show that if H and K are normal subgroups of G, then their intersection $H \cap K$ is also a normal subgroup.

4. Construct all normal subgroups of $I/(12)$ and for each normal subgroup, construct the Cayley table for the corresponding factor group.

5. Let H and K be normal subgroups of G such that $H \subseteq K$. Show that it is possible to construct a homomorphism $f: (G/H) \rightarrow (G/K)$. What is the kernel of f?

Exercise 5-10

1. Does there exist a homomorphism $f: I/(11) \rightarrow I/(3)$?

2. Does there exist a homomorphism $f: I/(6) \rightarrow I(3)$?

3. Let $f: G \rightarrow G'$ be a homomorphism, where G has order n. Show that if $a \in G$ has order t and $(t,n) = 1$, then a is contained in the kernel of f.

4. Let H and K be subgroups of G such that $H \lhd G$. Let

$$HK = \{hk \mid h \in H, k \in K\}$$

   (a) Show that HK is a subgroup of G.
   (b) Let f be the mapping $f: K \rightarrow (HK/H)$ defined by the rule

$$f(x) = Hx, \ x \in K$$

   Verify that f is a homomorphism whose kernel is $H \cap K$.
   (c) Prove that $(HK : 1) = (H : 1)(K : 1)/(H \cap K : 1)$.

5. An element a of an infinite group G is said to be *periodic* if $a^n = 1$ for some integer n.
   (a) Show that if G is an Abelian group, then the set P of its periodic elements is a subgroup of G.
   (b) Then show that the only periodic element of the factor group G/P is the identity.

6. Describe all homomorphic images of the group S_3.

Exercise 5-11

1. Using the Cayley Representation Theorem, represent the group $I/(6)$ as a subgroup of S_6.

2. Represent the group G_8 as a subgroup of S_4.

3. Find a subgroup of S_3 which is a homomorphic image of $I/(6)$.

4. Let G be a group of order 24 having a subgroup H of order 6 such that the only normal subgroup K of G which is contained in H is $K = \{1\}$. Prove that G is isomorphic to S_4.

Exercise 5-12

1. Find the class equation for S_3.

2. Find the class equation for the isometry group of the square.

3. Prove that the center Z of a group G is a normal subgroup of G.

4. Calculate the center of the group S_3.

5. Let G be a group of order p^n, where p is a prime. Using the class equation, show that the center of G has order larger than one.

6. Let G be a group of order p^n, where p is a prime and Z is its center. Show that if H is a normal subgroup of G such that $H \neq \{1\}$, then $H \cap Z \neq \{1\}$. (Hint: Use the result of the preceding problem, together with induction on n — or use the class equation.)

Exercise 5-13

1. Let G be a group of order 20 with the property that each of its subgroups is a normal subgroup. Prove that G is Abelian.

2. Determine all groups of order 14.

3. Let G be a group of order $p^\alpha m$, where p is a prime and $(p,m) = 1$. Show that for each integer $k \leq n$, G contains a subgroup of order p^k.

Exercise 5-14

1. Let Π be the affine geometry which is constructed from the field $I/(p)$, where p is a prime. Let A be the translation group of Π. Show that every element of A has order p.

2. The following is the multiplication table for an Abelian group A of order 9. Using the Cayley Representation Theorem, construct an affine geometry of 9 points and 12 lines which has A as its translation group.

	1	a	b	c	d	e	f	g	h
1	1	a	b	c	d	e	f	g	h
a	a	b	1	d	e	c	g	h	f
b	b	1	a	e	c	d	h	f	g
c	c	d	e	f	g	h	1	a	b
d	d	e	c	g	h	f	a	b	1
e	e	c	d	h	f	g	b	1	a
f	f	g	h	1	a	b	c	d	e
g	g	h	f	a	b	1	d	e	c
h	h	f	g	b	1	a	e	c	d

3. Let Π be an affine geometry for which Axiom 5-10 fails to be true. Prove that there exists no field which can coordinatize the points and lines of Π.

4. Let Π be the affine geometry consisting of the points B,D,E,F,G,H,J,K,L and the lines

   $l_1 = \{B,D,J\}$ $\quad l_2 = \{B,E,K\}$ $\quad l_3 = \{D,F,L\}$ $\quad l_4 = \{D,E,G\}$

   $l_5 = \{E,F,H\}$ $\quad l_6 = \{F,G,B\}$ $\quad l_7 = \{G,H,J\}$ $\quad l_8 = \{H,K,D\}$

   $l_9 = \{J,L,E\}$ $\quad l_{10} = \{J,K,F\}$ $\quad l_{11} = \{K,L,G\}$ $\quad l_{12} = \{L,B,H\}$

   Calculate the translation group A of this geometry by using just Properties 5-8 and 5-9. (Hint: Consider the subgroup of A fixing line l_1.)

6

The Classical Theory of Equations

6-1 Historical Note

The problem of solving algebraic equations of degree n occupied mathematicians for many centuries. While the ancient Greeks and Babylonians knew the solution of the quadratic equation, it was not until the seventeenth century that the cubic and quartic equations were solved by the Italian School of Mathematics.

The quintic equation resisted all efforts of solution, and by the time of the Age of Reason, whem mathematics was given a preeminent role, the problem of solving the fifth degree equation was one of the most famous in science.

In 1824, Niels Henrik Abel proved the impossibility of solving the quintic equation by radicals. Shortly after, Evariste Galois showed exactly which equations of degree n could be solved by radicals.*

These papers of Abel and Galois, in which the theory of groups was used to settle the solvability of algebraic equations, are generally regarded as the beginnings of modern algebra.

The basis for attacking the solution of algebraic equations with groups was set forth in 1770 by Joseph Lagrange in a beautiful memoir, *Réflexions sur la résolution algébraique des équations.* In that memoir Lagrange catalogued the various techniques for solving algebraic equations and pointed out that these results could be described most succinctly in terms of permutations of the roots of the equations. Near the end of his paper Lagrange said,

> "Here, if I am not mistaken, are the true principles for the solution of equations and the most appropriate analysis to lead to them; all is reduced to a kind of calculus of combinations by which one finds *a priori* the results one should expect."†

Lagrange's memoir made a deep impression upon Abel and Galois.

In this chapter we shall give Lagrange's theory of equations. This can be used as a basis for the more extensive *Galois Theory*, which is adequately described in a number of books. We particularly recommend B. L. van der Waerden's *Modern Algebra*, vol 1 (Frederick Ungar Publishing Co., New York 1949) and Emil Artin's *Galois Theory* (Notre Dame Mathematical Lectures, Number 2; University of Notre Dame; Notre Dame, Indiana 1942) for an account of this theory.

6-2 Symmetric Functions

A *symmetric* function is a polynomial $f = f(x_1, \ldots, x_n)$ in variables $x_1, \ldots, x_n$ such that $f(x_1, \ldots, x_n) = f(x_{\sigma(1)}, \ldots, x_{\sigma(n)})$ for every permutation σ of the integers $1, \ldots, n$. For example, $(x_1 + x_2 + x_3)^4 + (x_1^2 + x_2^2 + x_3^2)^3$ is left unchanged by every permutation of the integers 1,2,3; hence, it is a symmetric function of x_1, x_2, and x_3. On the other

---

*In spite of its importance, many years passed before Galois' work became widely known.

†*Oeuvres de Lagrange*, vol 3, p. 403. Gauthier-Villars, Paris.

hand, the function $f(x_1,x_2,x_3) = x_1x_2 + x_1x_3$ is not symmetric since for the permutation

$$\sigma = \begin{pmatrix} 1 & 2 & 3 \\ 2 & 3 & 1 \end{pmatrix}$$

$f(x_{\sigma(1)},x_{\sigma(2)},x_{\sigma(3)}) = x_2x_3 + x_2x_1 \neq f(x_1,x_2,x_3)$.

For a field F the *elementary* symmetric functions in $F[x_1, \ldots, x_n]$ are defined to be

$$\begin{aligned} s_1 &= x_1 + \ldots + x_n \\ s_2 &= \sum_{1 \leq i < j \leq n} x_i x_j \\ s_3 &= \sum_{1 \leq i < j < k \leq n} x_i x_j x_k \\ &\cdot \\ &\cdot \\ &\cdot \\ s_n &= x_1 x_2 \ldots x_n \end{aligned} \tag{6-1}$$

The elementary symmetric functions in the case $n = 3$ are

$$\begin{aligned} s_1 &= x_1 + x_2 + x_3 \\ s_2 &= x_1x_2 + x_1x_3 + x_2x_3 \\ s_3 &= x_1x_2x_3 \end{aligned}$$

The importance of elementary symmetric functions to the theory of equations lies in the following fact: If $f(x) = x^n + a_{n-1}x^{n-1} + \ldots + a_0$ is a polynomial of degree n with roots $\alpha_1, \ldots, \alpha_n$, then the coefficients of $f(x)$ are expressible in terms of the elementary symmetric functions of $\alpha_1, \ldots, \alpha_n$. Indeed, factoring $f(x) = (x - \alpha_1) \ldots (x - \alpha_n)$ and comparing coefficients, we have

$$\begin{aligned} a_{n-1} &= -(\alpha_1 + \ldots + \alpha_n) = -s_1(\alpha_1, \ldots, \alpha_n) \\ a_{n-2} &= \sum_{1 \leq i < j \leq n} \alpha_i \alpha_j = s_2(\alpha_1, \ldots, \alpha_n) \\ &\cdot \\ &\cdot \\ &\cdot \\ a_0 &= (-1)^n \alpha_1 \ldots \alpha_n = (-1)^n s_n(\alpha_1, \ldots, \alpha_n) \end{aligned} \tag{6-2}$$

For a given polynomial $f(x) = x^n + a_{n-1}x^{n-1} + \ldots + a_0$, the problem of finding its roots $\alpha_1, \ldots, \alpha_n$ is equivalent to solving the system (6-2) for $\alpha_1, \ldots, \alpha_n$. If $f(x) = x^2 + a_1x + a_0$ is quadratic, the equations

$$-a_1 = \alpha_1 + \alpha_2$$
$$a_0 = \alpha_1\alpha_2$$

are readily solved, giving the quadratic formulas

$$\alpha_1 = -\frac{1}{2}a_1 + \frac{1}{2}\sqrt{D}$$
$$\alpha_2 = -\frac{1}{2}a_1 - \frac{1}{2}\sqrt{D}$$

where $D = a_1^2 - 4a_0$.

The following result is known as the Fundamental Theorem on symmetric functions:

Theorem 6-1. Every symmetric function in $F[x_1, \ldots, x_n]$ can be written as a polynomial in the elementary symmetric functions.

Proof: Any polynomial in $F[x_1, \ldots, x_n]$ is composed of terms of the form

$$A = x_1^{e_1}x_2^{e_2} \ldots x_n^{e_n} \quad \textbf{(6-3)}$$

where $e_1, e_2, \ldots, e_n$ are nonnegative integers. We will want to compare a term of the form (6-3) with one of the form

$$B = x_1^{f_1}x_2^{f_2} \ldots x_n^{f_n} \quad \textbf{(6-4)}$$

where $f_1, f_2, \ldots, f_n$ are again nonnegative integers. Now if $A \neq B$, then for some i, $e_i \neq f_i$. Suppose t is the smallest integer for which $e_t \neq f_t$. Then we shall say that A is *larger* than B if $e_t > f_t$ and call B larger than A if $e_t < f_t$. For example, the term $x_1^3x_2^3x_3^5x_4^2x_5$ is larger than $x_1^3x_2^3x_3^4x_4^9x_5^8$.

Suppose now that $g(x_1, \ldots, x_n)$ is a symmetric function of $x_1, \ldots, x_n$ and let

$$ax_1^{e_1}x_2^{e_2} \ldots x_n^{e_n} \quad \textbf{(6-5)}$$

be its largest term, where $a \in F$. As $g(x_1, \ldots, x_n)$ is a symmetric function, it must contain among its terms

$$ax_{\sigma(1)}^{e_1}x_{\sigma(2)}^{e_2} \ldots x_{\sigma(n)}^{e_n}$$

for every permutation σ of $1,2, \ldots, n$. Therefore, for the term (6-5), we must have

$$e_1 \geq e_2 \geq \ldots \geq e_n$$

Now consider the polynomial

$$as_1^{e_1-e_2} s_2^{e_2-e_3} \dots s_{n-1}^{e_{n-1}-e_n} s_n^{e_n} \tag{6-6}$$

Evidently, the largest term of (6-6) is

$$ax_1^{e_1-e_2}(x_1x_2)^{e_2-e_3} \dots (x_1x_2 \dots x_{n-1})^{e_{n-1}-e_n}(x_1x_2 \dots x_n)^{e_n}$$

and this reduces to (6-5). Therefore the term (6-5) does not appear in the polynomial

$$h(x_1, \dots, x_n) = g(x_1, \dots, x_n) - as_1^{e_1-e_2}s_2^{e_2-e_3} \dots s_n^{e_n}$$

In other words, the largest term of $h(x_1, \dots, x_n)$ is smaller than (6-5). Moreover $h(x_1, \dots, x_n)$ is a symmetric function. Altogether then

$$g(x_1, \dots, x_n) = as_1^{e_1-e_2}s_2^{e_2-e_3} \dots s_n^{e_n} + h(x_1, \dots, x_n)$$

where $h(x_1, \dots, x_n)$ is a symmetric function whose largest term is smaller than the largest term of $g(x_1, \dots, x_n)$. Eliminating next the largest term of $h(x_1, \dots, x_n)$, and so on, eventually we will express $g(x_1, \dots, x_n)$ as a polynomial in the elementary symmetric functions $s_1, s_2, \dots, s_n$.

Remark 6-1. In practice, to express a symmetric function as a polynomial in $s_1,s_2, \dots, s_n$ we may proceed as in the proof of Theorem 6-1. For example, the largest term of $g(x_1,x_2,x_3) = x_1^2 + x_2^2 + x_3^2$ is $x_1^2x_2^0x_3^0$. Then $g(x_1,x_2,x_3) - s_1^2 = x_1^2 + x_2^2 + x_3^2 - (x_1 + x_2 + x_3)^2 = -2(x_1x_2 + x_1x_3 + x_2x_3)$; hence, $g(x_1,x_2,x_3) = s_1^2 - 2s_2$.

6-3 The Discriminant of a Polynomial

The roots x_1, x_2 of the quadratic

$$f(x) = x^2 + px + q$$

are given by $x_1 = -1/2p + 1/2\sqrt{D}$, $x_2 = -1/2p + \sqrt{D}$, where the discriminant $D = (x_1 - x_2)^2$.

In general, if $f(x)$ is a polynomial of degree n with roots $x_1,x_2, \dots, x_n$, then its *discriminant* D is defined to be

$$D = \prod_{i<j} (x_i - x_j)^2$$

The discriminant D is a symmetric function of the roots $x_1, \dots, x_n$ and can be expressed as a polynomial in the elementary symmetric functions of $x_1, \ldots, x_n$ because of the Fundamental Theorem on symmetric functions. Using the equations (6-2), we may write D in terms of the coefficients of the polynomial $f(x)$.

For the cubic

$$f(x) = x^3 + px + q \tag{6-7}$$

the discriminant is

$$D = (x_1 - x_2)^2(x_1 - x_3)^2(x_2 - x_3)^2$$

where x_1,x_2,x_3 denote the roots of $f(x)$. Since

$$f(x) = (x - x_1)(x - x_2)(x - x_3) \tag{6-8}$$

we have

$$\begin{aligned} s_1 &= x_1 + x_2 + x_3 = 0 \\ s_2 &= x_1x_2 + x_1x_3 + x_2x_3 = p \\ s_3 &= x_1x_2x_3 = -q \end{aligned} \tag{6-9}$$

Now to write D as a polynomial in s_1,s_2,s_3, we may proceed as in Remark 6-1. However, it is simpler to observe that from Equation (6-8)

$$-D = f'(x_1)f'(x_2)f'(x_3)$$

Calculating the derivative from Equation (6-7), we have

$$\begin{aligned} -D = (3x_1^2 + p)(3x_2^2 + p)(3x_3^2 + p) &= 27x_1^2x_2^2x_3^2 \\ &+ 9p(x_1^2x_2^2 + x_1^2x_3^2 + x_2^2x_3^2) + 3p^2(x_1^2 + x_2^2 + x_3^2) + p^3 \end{aligned}$$

Now,

$$x_1^2x_2^2x_3^2 = s_3^2 = q^2$$

and

$$x_1^2 + x_2^2 + x_3^2 = s_1^2 - 2s_2 = -2p$$

Thus,

$$-D = 27q^2 + 9p(x_1^2x_2^2 + x_1^2x_3^2 + x_2^2x_3^2) - 6p^3 + p^3$$

What remains is to write

$$x_1^2x_2^2 + x_1^2x_3^2 + x_2^2x_3^2$$

in terms of the elementary symmetric functions. However, this term is

$$\begin{aligned} &(x_1x_2 + x_1x_3 + x_2x_3)^2 - 2(x_1^2x_2x_3 + x_2^2x_1x_3 + x_3^2x_1x_2) \\ &\qquad = s_2^2 - 2(x_1 + x_2 + x_3)(x_1x_2x_3) = s_2^2 - 2s_1s_3 = p^2 \end{aligned}$$

Therefore,

$$-D = 27q^2 + 9p^3 - 6p^3 + p^3$$

Altogether, we have proved the following theorem:

Theorem 6-2. The discriminant D of the cubic

$$x^3 + px + q$$

is $D = -27q^2 - 4p^3$.

6-4 The Subgroup Fixing a Polynomial

We investigate now those polynomials of $F[x_1, \ldots, x_n]$ which are not symmetric. For a polynomial $f = f(x_1, \ldots, x_n)$ and a permutation $\sigma \in S_n$, let

$$f_\sigma = f(x_{\sigma(1)}, \ldots, x_{\sigma(n)})$$

For example, if $f(x_1,x_2,x_3) = x_1x_3 + x_2^4$ and

$$\sigma = \begin{pmatrix} 1 & 2 & 3 \\ 2 & 3 & 1 \end{pmatrix}$$

then $f_\sigma = x_2x_1 + x_3^4$.

The set H of all $\sigma \in S_n$ for which $f = f_\sigma$ is a subgroup of S_n, and we call *H the subgroup fixing f.* In the preceding example, $f = f_\sigma$ for the permutations $\sigma = 1$ and

$$\sigma = \begin{pmatrix} 1 & 2 & 3 \\ 3 & 2 & 1 \end{pmatrix}$$

Thus,

$$H = \left\{1, \begin{pmatrix} 1 & 2 & 3 \\ 3 & 2 & 1 \end{pmatrix}\right\}$$

is the subgroup fixing f.

For a function f of $F[x_1, \ldots, x_n]$ the functions f_σ, as σ ranges over the entire group S_n, can be determined once the subgroup H fixing f is known, For if

$$S_n = H \cup H\sigma_2 \cup \ldots \cup H\sigma_m$$

is the coset decomposition of the group S_n relative to the subgroup H, then the functions

$$f, f_{\sigma_2}, \ldots, f_{\sigma_m} \qquad \textbf{(6-10)}$$

are all distinct. Furthermore, if σ is any element of S_n, then the cosets

$$H\sigma, H\sigma_2\sigma, \ldots, H\sigma_m\sigma$$

are all distinct; hence,

$$S_n = H\sigma \cup H\sigma_2\sigma \cup \ldots \cup H\sigma_m\sigma$$

In particular, $H\sigma = H\sigma_i$ for some i, whence, $f_\sigma = f_{\sigma_i}$. It follows that the functions (6-10) are all possible values f_σ, for $\sigma \in S_n$.

From this observation we have immediately the following interesting result:

Theorem 6-3. Let H be the subgroup of S_n fixing the function $f(x_1, \ldots, x_n)$ and let

$$S_n = H \cup H\sigma_2 \cup \ldots \cup H\sigma_m$$

be the coset decomposition of S_n relative to H. Then the coefficients of the polynomial

$$G(t) = (t - f)(t - f_{\sigma_2}) \ldots (t - f_{\sigma_m})$$

are symmetric functions of $x_1, \ldots, x_n$.

Proof: For applying a permutation σ to a typical coefficient

$$s_i(f, f_{\sigma_2}, \ldots, f_{\sigma_m})$$

of $G(t)$, we find $s_i(f, f_{\sigma_2}, \ldots, f_{\sigma_m})\sigma = s_i(f_\sigma, f_{\sigma_2\sigma}, \ldots, f_{\sigma_m\sigma}) = s_i(f, f_{\sigma_2}, \ldots, f_{\sigma_m})$, since σ simply permutes $f, f_{\sigma_2}, \ldots, f_{\sigma_m}$ and s_i is symmetric.

Evidently, if H is the subgroup fixing a polynomial f than H fixes any polynomial of the form

$$p_0 + p_1 f + \ldots + p_k f^k \qquad \textbf{(6-11)}$$

where $p_0, p_1, \ldots, p_k$ are symmetric functions. Furthermore, H will fix the ratio of two expressions such as (6-11). That is, H leaves unchanged any expression of the form

$$\frac{A(f)}{B(f)} \qquad \textbf{(6-12)}$$

where $A(t)$ and $B(t)$ are polynomials in t whose coefficients are symmetric functions. The functions (6-12) turn out to be the only ones fixed by H, as is shown in the following theorem:

Theorem 6-4. Let H be the subgroup fixing a polynomial $f(x_1, \dots, x_n)$ of $F[x_1, \dots, x_n]$. If $g(x_1, \dots, x_n)$ is also fixed by H, i.e., $g = g_\sigma$ for all σ in H, then

$$g = \frac{A(f)}{B(f)}$$

for some polynomials $A(t)$ and $B(t)$, whose coefficients are symmetric functions of $x_1, \dots, x_n$.

Proof: Decomposing the group S_n into cosets relative to H, we have

$$S_n = H \cup H\sigma_2 \cup \dots \cup H\sigma_m$$

for some $\sigma_2, \dots, \sigma_m$ in S_n. Then the polynomials

$$C(t) = (t - f)(t - f_{\sigma_2}) \dots (t - f_{\sigma_m})$$

and

$$A(t) = (t - f)(t - f_{\sigma_2}) \dots (x - f_{\sigma_m})\left[\frac{g}{t - f} + \frac{g_{\sigma_2}}{t - f_{\sigma_2}} + \dots + \frac{g_{\sigma_m}}{t - f_{\sigma_m}}\right]$$

have coefficients which are symmetric functions in $x_1, \dots, x_n$. Setting $t = f$ in $A(t)$, we have

$$A(f) = (t - f_{\sigma_2}) \dots (f - f_{\sigma_m})g = C'(f)g$$

where $C' = dC/dt$. Thus,

$$g = \frac{A(f)}{B(f)}$$

where we have set $B(t) = C'(t)$.

6-5 The Solution of Quartic Equations

We wish to show now how equations may be solved by group-theoretic methods. In order to put aside certain technical difficulties, we shall assume that it is already known how to solve cubic equations.

To solve the quartic $g(x) = x^4 + ax^3 + bx^2 + cx + d = 0$, it will be convenient to eliminate the x^3 term from the quartic. This is done by setting $f(x) = g(x - (a/4))$. The resulting polynomial $f(x)$ will have the form

$$f(x) = x^4 + px^2 + qx + r \qquad \textbf{(6-13)}$$

If the roots of $f(x)$ are $\alpha_1,\alpha_2,\alpha_3,\alpha_4$, then the roots of $g(x)$ will be $\alpha_i - (a/4), i = 1,2,3,4$.

We have not specified yet in which field F the coefficient of $f(x)$ lie. All that is required is for F to have characteristic different from 2 or 3. (The restriction that the characteristic be different from 2 was already necessary in eliminating the x^3 term from the quartic — for other computations involving a cubic polynomial one requires the characteristic different from 3 as well.) However, to fix ideas, we shall take F to be the complex field.

For the roots $\alpha_1,\alpha_2,\alpha_3,\alpha_4$ of $f(x)$ we have the factorization

$$f(x) = (x - \alpha_1)(x - \alpha_2)(x - \alpha_3)(x - \alpha_4) \qquad \textbf{(6-14)}$$

hence, upon comparing coefficients in Equations (6-13) and (6-14)

$$\begin{aligned} \alpha_1 + \alpha_2 + \alpha_3 + \alpha_4 &= 0 \\ s_2(\alpha_1,\alpha_2,\alpha_3,\alpha_4) &= p \\ s_3(\alpha_1,\alpha_2,\alpha_3,\alpha_4) &= -q \\ \alpha_1\alpha_2\alpha_3\alpha_4 &= r \end{aligned} \qquad \textbf{(6-15)}$$

where $s_2(\alpha_1,\alpha_2,\alpha_3,\alpha_4)$ and $s_3(\alpha_1,\alpha_2,\alpha_3,\alpha_4)$ denote elementary symmetric functions of the roots $\alpha_1,\alpha_2,\alpha_3,\alpha_4$.

The quantity

$$u = (\alpha_1 + \alpha_2)(\alpha_3 + \alpha_4)$$

is fixed by a subgroup H of the symmetric group S_4. This subgroup H has order 8 (see Exercise 6-4, Problem 2). Decomposing S_4 into distinct cosets relative to H,

$$S_4 = H \cup H\lambda \cup H\mu, \ \lambda,\mu \in S_4$$

It follows that for any $\sigma \in S_4$, $u_\sigma = u, u_\lambda$, or u_μ. In other words, there are exactly three possible results for u_σ, $\sigma \in S_4$. Evidently these are

$$\begin{aligned} u &= (\alpha_1 + \alpha_2)(\alpha_3 + \alpha_4) \\ u_\lambda &= (\alpha_1 + \alpha_3)(\alpha_2 + \alpha_4) \\ u_\mu &= (\alpha_1 + \alpha_4)(\alpha_2 + \alpha_3) \end{aligned} \qquad \textbf{(6-16)}$$

where

$$\lambda = \begin{pmatrix} 1 & 2 & 3 & 4 \\ 1 & 3 & 2 & 4 \end{pmatrix}$$
$$\mu = \begin{pmatrix} 1 & 2 & 3 & 4 \\ 1 & 4 & 2 & 3 \end{pmatrix}$$

The polynomial

$$G(x) = (x - u)(x - u_\lambda)(x - u_\mu) \qquad \textbf{(6-17)}$$

has coefficients which are symmetric functions of $\alpha_1,\alpha_2,\alpha_3,\alpha_4$ (see Theorem 6-3). Using the Fundamental Theorem on symmetric functions, these coefficients can be made to depend on the elementary symmetric functions of $\alpha_1,\alpha_2,\alpha_3,\alpha_4$, and these in turn may be written in terms of p,q, and r by virtue of Equations (6-15). In fact, multiplying out the right-hand side of Equation (6-17), we have $G(x) = x^3 - (u + u_\lambda + u_\mu)x^2 + (uu_\lambda + uu_\mu + u_\lambda u_\mu)x - uu_\lambda u_\mu$. From Equation (6-16), $u + u_\lambda + u_\mu = 2(\alpha_1\alpha_2 + \alpha_1\alpha_3 + \alpha_1\alpha_4 + \alpha_2\alpha_3 + \alpha_2\alpha_4 + \alpha_3\alpha_4) = 2s_2(\alpha_1,\alpha_2,\alpha_3,\alpha_4) = 2p$ because of (6-15). In a similar way, after lengthy calculation, $uu_\lambda + uu_\mu + u_\lambda u_\mu = s_2^2 + s_1s_3 - 4s_4$ and $uu_\lambda u_\mu = s_1s_2s_3 - s_1^2s_4 - s_3^2$. Hence, by (6-15)

$$G(x) = x^3 - 2px^2 + (p^2 - 4r)x + q^2 \tag{6-18}$$

Now, we have already assumed that cubics can be solved. Hence, if θ_1, θ_2, and θ_3 are found to be the roots of $G(x)$, then

$$\begin{aligned}(\alpha_1 + \alpha_2)(\alpha_3 + \alpha_4) &= \theta_1\\ (\alpha_1 + \alpha_3)(\alpha_2 + \alpha_4) &= \theta_2\\ (\alpha_1 + \alpha_4)(\alpha_2 + \alpha_3) &= \theta_3\end{aligned} \tag{6-19}$$

As $\alpha_1 + \alpha_2 + \alpha_3 + \alpha_4 = 0$, we obtain from Equations (6-19)

$$\begin{aligned}\alpha_1 + \alpha_2 &= \sqrt{-\theta_1}\\ \alpha_1 + \alpha_3 &= \sqrt{-\theta_2}\\ \alpha_1 + \alpha_4 &= \sqrt{-\theta_3}\end{aligned} \tag{6-20}$$

Adding these equations and using the relation $\alpha_1 + \alpha_2 + \alpha_3 + \alpha_4 = 0$, we find

$$\begin{aligned}\alpha_1 &= \frac{1}{2}(\sqrt{-\theta_1} + \sqrt{-\theta_2} + \sqrt{-\theta_3})\\ \alpha_2 &= \frac{1}{2}(\sqrt{-\theta_1} - \sqrt{-\theta_2} - \sqrt{-\theta_3})\\ \alpha_3 &= \frac{1}{2}(-\sqrt{-\theta_1} + \sqrt{-\theta_2} - \sqrt{-\theta_3})\\ \alpha_4 &= \frac{1}{2}(-\sqrt{-\theta_1} - \sqrt{-\theta_2} + \sqrt{-\theta_3})\end{aligned} \tag{6-21}$$

We have expressed the solutions of the quartic $f(x) = 0$ in terms of the solutions of a cubic equation. Note that in going from Equation (6-19) to Equation (6-20) we have not explained which of the two possible square roots are to be taken from $-\theta_1, -\theta_2$, and $-\theta_3$. This is determined by actually substituting into the equation $f(x) = 0$ the values of Equation (6-21) we have found for $\alpha_1,\alpha_2,\alpha_3$, and α_4.

6-6 The Solution of Cubic Equations

To solve the cubic $g(x) = x^3 + ax^2 + bx + c$ we eliminate, as before, the x^2 term by setting $f(x) = g(x - (a/3))$. The resulting polynomial is

$$f(x) = x^3 + px + q$$

where $p = b - (a^2/3)$ and $q = c + (2a^3 - 9ab)/27$. The field in which the coefficients of $f(x)$ lie is taken to be the complex field.

We have the factorization

$$f(x) = (x - \alpha_1)(x - \alpha_2)(x - \alpha_3) \tag{6-22}$$

where α_1, α_2, and α_3 are the roots of $f(x)$; hence,

$$\begin{aligned} \alpha_1 + \alpha_2 + \alpha_3 &= 0 \\ \alpha_1\alpha_2 + \alpha_1\alpha_3 + \alpha_2\alpha_3 &= p \\ \alpha_1\alpha_2\alpha_3 &= -q \end{aligned} \tag{6-23}$$

Now to find the roots of $f(x)$ it is sufficient to solve the system (6-23) for $\alpha_1, \alpha_2, \alpha_3$. However, as the last two equations of (6-23) are nonlinear in the roots, it is better to form first the *Lagrange resolvents*. These are the quantities

$$\begin{aligned} \theta &= \alpha_1 + \zeta\alpha_2 + \zeta^2\alpha_3 \\ \psi &= \alpha_1 + \zeta^2\alpha_2 + \zeta\alpha_3 \end{aligned} \tag{6-24}$$

where $\zeta = \cos(2\pi/3) + i\sin(2\pi/3)$ is a primitive cube root of unity. For ζ we have

$$\zeta^3 = 1 \quad \text{and} \quad 1 + \zeta + \zeta^2 = 0 \tag{6-25}$$

Using the relations of Equation (6-25) the roots $\alpha_1, \alpha_2, \alpha_3$ may be expressed quite nicely in terms of the Lagrange resolvents. For we have the system

$$\begin{aligned} \alpha_1 + \alpha_2 + \alpha_3 &= 0 \\ \alpha_1 + \zeta\alpha_2 + \zeta^2\alpha_3 &= \theta \\ \alpha_1 + \zeta^2\alpha_2 + \zeta\alpha_3 &= \psi \end{aligned} \tag{6-26}$$

Adding these equations, and using the second relation of Equation (6-25) we find $\alpha_1 = (1/3)(\theta + \psi)$. By multiplying the second equation of (6-26) by ζ^2, the third equation by ζ and adding, $\alpha_2 = (1/3)(\zeta^2\theta + \zeta\psi)$. Similarly, $\alpha_3 = (1/3)(\zeta\theta + \zeta^2\psi)$. Hence,

$$\begin{aligned} \alpha_1 &= \frac{1}{3}(\theta + \psi) \\ \alpha_2 &= \frac{1}{3}(\zeta^2\theta + \zeta\psi) \\ \alpha_3 &= \frac{1}{3}(\zeta\theta + \zeta^2\psi) \end{aligned} \tag{6-27}$$

To find the roots of $f(x)$ we must calculate the Lagrange resolvents θ and ψ, and to do this, we form the quantity

$$\Delta = (\alpha_1 - \alpha_2)(\alpha_1 - \alpha_3)(\alpha_2 - \alpha_3) \tag{6-28}$$

Considering Δ as a function of the roots $\alpha_1,\alpha_2,\alpha_3$, let H be the subgroup of S_3 fixing Δ. Evidently

$$H = \left\{1, \sigma = \begin{pmatrix} 1 & 2 & 3 \\ 2 & 3 & 1 \end{pmatrix}, \sigma^2 = \begin{pmatrix} 1 & 2 & 3 \\ 3 & 1 & 2 \end{pmatrix}\right\} \tag{6-29}$$

Moreover,

$$\Delta^2 = D, \quad \text{the discriminant} \tag{6-30}$$

From an earlier calculation (see Theorem 6-2)

$$D = -27q^2 - 4p^3 \tag{6-31}$$

Now for the Lagrange resolvent θ note that the product $\theta\theta_\sigma\theta_{\sigma^2}$ is fixed by the subgroup H. Indeed,

$$\sigma : \theta\theta_\sigma\theta_{\sigma^2} \longrightarrow \theta_\sigma\theta_{\sigma^2}\theta_{\sigma^3}$$

But $\sigma^3 = 1$, whence, $\theta_\sigma\theta_{\sigma^2}\theta_{\sigma^3} = \theta\theta_\sigma\theta_{\sigma^2}$. Similarly, σ^2 fixes $\theta\theta_\sigma\theta_{\sigma^2}$. Applying Theorem 6-4, it follows that

$$\theta\theta_\sigma\theta_{\sigma^2} = \frac{A(\Delta)}{B(\Delta)} \tag{6-32}$$

for polynomials $A(t)$ and $B(t)$, whose coefficients are symmetric functions in $\alpha_1,\alpha_2,\alpha_3$. But because of the Fundamental Theorem on symmetric functions and the relations of Equation (6-23), we may claim further that $A(t)$ and $B(t)$ belong to $F[t]$.

Equation (6-32) may be reduced further. In the first place, $\Delta^2 = -27q^2 - 4p^3$ because of Equations (6-30) and (6-31). Hence, the denominator $B(\Delta) = b_0 + b_1\Delta + b_2\Delta^2 + \ldots$ reduces to $c + d\Delta$, where the quantities c and d can be expressed in terms of the coefficients of the cubic. Similarly, $A(\Delta) = a + b\Delta$, where a and b can be likewise expressed. Then rationalizing the denominator in the expression

$$\frac{a + b\Delta}{c + d\Delta}$$

we have

$$\theta\theta_\sigma\theta_{\sigma^2} = u + v\Delta \tag{6-33}$$

Remarkably, $\theta\theta_\sigma\theta_{\sigma^2}$ also simplifies (and this is another feature of the Lagrange resolvents). Indeed, as

$$\sigma = \begin{pmatrix} 1 & 2 & 3 \\ 2 & 3 & 1 \end{pmatrix}$$

and $\theta = \alpha_1 + \zeta\alpha_2 + \zeta^2\alpha_3$, $\theta_\sigma = \alpha_2 + \zeta\alpha_3 + \zeta^2\alpha_1 = \zeta^2\theta$. Also, $\theta_{\sigma^2} = \alpha_3 + \zeta\alpha_1 + \zeta^2\alpha_2 = \zeta\theta$. Hence, the product $\theta\theta_\sigma\theta_{\sigma^2} = \theta(\zeta^2\theta)(\zeta\theta) = \zeta^3\theta^3 = \theta^3$. Therefore, from Equation (6-33) we have

$$\theta^3 = u + v\Delta \tag{6-34}$$

Similarly, for the resolvent ψ,

$$\psi^3 = u' + v'\Delta \tag{6-35}$$

where u' and v' may be expressed in terms of the coefficients of the cubic equation.

Taking cube roots in Equations (6-34) and (6-35) and going back to Equation (6-27), we find

$$\begin{aligned} \alpha_1 &= \frac{1}{3}\sqrt[3]{u + v\sqrt{D}} + \frac{1}{3}\sqrt[3]{u' + v'\sqrt{D}} \\ \alpha_2 &= \frac{1}{3}\zeta^2\sqrt[3]{u + v\sqrt{D}} + \frac{1}{3}\zeta\sqrt[3]{u' + v'\sqrt{D}} \\ \alpha_3 &= \frac{1}{3}\zeta\sqrt[3]{u + v\sqrt{D}} + \frac{1}{3}\zeta^2\sqrt[3]{u' + v'\sqrt{D}} \end{aligned} \tag{6-36}$$

Of course, for the formulas (6-36) to make any sense we must calculate u,v,u',v' and decide which cube roots are to be taken in $\sqrt[3]{u + v\sqrt{D}}$ and $\sqrt[3]{u' + v'\sqrt{D}}$. For the moment we remark that it is in the sense of (6-36) that the cubic was said to be solvable by radicals in the time of Lagrange.

We now calculate that

$$\begin{aligned} \theta^3 = (\alpha_1 + \zeta\alpha_2 + \zeta^2\alpha_3)^3 &= (\alpha_1^3 + \alpha_2^3 + \alpha_3^3 + 6\alpha_1\alpha_2\alpha_3) \\ &+ 3\zeta(\alpha_1^2\alpha_2 + \alpha_2^2\alpha_3 + \alpha_3^2\alpha_1) + 3\zeta^2(\alpha_1\alpha_2^2 + \alpha_2\alpha_3^2 + \alpha_3\alpha_1^2) \end{aligned}$$

Moreover, $\Delta = P - Q$, where

$$P = \alpha_1^2\alpha_2 + \alpha_2^2\alpha_3 + \alpha_3^2\alpha_1$$

and

$$Q = \alpha_1\alpha_2^2 + \alpha_2\alpha_3^2 + \alpha_3\alpha_1^2$$

Therefore, $\theta^3 = (\alpha_1^3 + \alpha_2^3 + \alpha_3^3 + 6\alpha_1\alpha_2\alpha_3) + 3\zeta P + 3\zeta^2 Q$. As $\zeta = -(1/2) + (1/2)\sqrt{3}\, i$ and $\zeta^2 = -(1/2) - (1/2)\sqrt{3}\, i$, we get

$$\theta^3 = (\alpha_1^3 + \alpha_2^3 + \alpha_3^3 + 6\alpha_1\alpha_2\alpha_3) - \frac{3}{2}(P + Q) + \frac{3}{2}\sqrt{3}i(P - Q)$$

or

$$\theta^3 = (\alpha_1^3 + \alpha_2^3 + \alpha_3^3 + 6\alpha_1\alpha_2\alpha_3) + \frac{3}{2}\sqrt{3}i\Delta - \frac{3}{2}(P + Q) \qquad \textbf{(6-37)}$$

The quantities $\alpha_1^3 + \alpha_2^3 + \alpha_3^3 + 6\alpha_1\alpha_2\alpha_3$ and $P + Q$ are symmetric functions in $\alpha_1,\alpha_2,\alpha_3$. Expressing them in terms of the elementary symmetric functions and using Equation (6-23) we get from Equation (6-37) the result

$$\theta^3 = -\frac{27}{2}q + \frac{3}{2}\sqrt{3}i\Delta \qquad \textbf{(6-38)}$$

Now to calculate ψ^3 observe that if

$$\tau = \begin{pmatrix} 1 & 2 & 3 \\ 1 & 3 & 2 \end{pmatrix}$$

Then $\tau : \Delta \to -\Delta$ and $\tau : \theta \to \psi$. Hence, applying the permutation τ to both sides of Equation (6-38) we find

$$\psi^3 = -\frac{27}{2}q - \frac{3}{2}\sqrt{3}i\Delta \qquad \textbf{(6-39)}$$

Then

$$\begin{aligned} \theta &= \sqrt[3]{-\frac{27}{2}q + \frac{3}{2}\sqrt{3}i\sqrt{D}} \\ \psi &= \sqrt[3]{-\frac{27}{2}q - \frac{3}{2}\sqrt{3}i\sqrt{D}} \end{aligned} \qquad \textbf{(6-40)}$$

In Equation (6-40) it has to be explained which of the three possible values one takes for the cube roots. For this note that

$$\begin{aligned} \theta\psi &= (\alpha_1 + \zeta\alpha_2 + \zeta^2\alpha_3)(\alpha_1 + \zeta^2\alpha_2 + \zeta\alpha_3) \\ &= (\alpha_1^2 + \alpha_2^2 + \alpha_3^2) + (\zeta + \zeta^3)(\alpha_1\alpha_3 + \alpha_1\alpha_2 + \alpha_2\alpha_3) \\ &= (\alpha_1 + \alpha_2 + \alpha_3)^2 + (-2 + \zeta + \zeta^2)(\alpha_1\alpha_3 + \alpha_1\alpha_2 + \alpha_2\alpha_3) \\ &= -3(\alpha_1\alpha_3 + \alpha_1\alpha_2 + \alpha_2\alpha_3) = -3p \end{aligned}$$

Therefore, in Equation (6-40) the cube roots must be taken so that

$$\theta\psi = -3p \tag{6-41}$$

Exercises

Exercise 6-1

1. Show that the cubic equation $x^3 + px + q = 0$ may be solved by introducing new variables y and z where $x = y + z$ and $3yz + p = 0$.

2. Show that for a quartic equation $x^4 + px^2 + qx + r = 0$ one can always find quantities a,b,c,d such that

$$x^4 + px^2 + qx + r = (x^2 + ax + b)(x^2 + cx + d)$$

Hence, in this way, obtain the solutions of the original quartic equation.

Exercise 6-2

1. Express $x_1^3 + x_2^3 + x_3^3$ as a polynomial in the elementary symmetric functions.

2. Express $x_1^2x_2^2 + x_1^2x_3^2 + x_2^2x_3^2$ as a polynomial in the elementary symmetric functions.

3. Let $f(x) = x^n + a_1x^{n-1} + \ldots + a_0$ be a polynomial with coefficients lying in some field F. Suppose the roots of $f(x)$ are $x_1, \ldots, x_n$. Show that if $g(x)$ is any polynomial in $F[x]$, then the polynomial

$$h(x) = (x - g(x_1)) \ldots (x - g(x_n))$$

belongs to $F[x]$.

Exercise 6-4

1. Find the order of the subgroup of S_6 fixing

$$x_1 + x_2 + x_3 - 2(x_4 + x_5 + x_6)$$

2. Find the order of the subgroup of S_4 fixing

$$(x_1 + x_2)(x_3 + x_4)$$

3. Let H be the subgroup of S_n fixing $f = f(x_1, \ldots, x_n)$. Show that if H is a normal subgroup of S_n, then for any σ in S_n,

$$f_\sigma = \frac{A(f)}{B(f)}$$

for some polynomials $A(t),B(t)$ whose coefficients are symmetric functions in $x_1, \ldots, x_n$.

4. Find a function $f(x_1, \ldots, x_n)$ of $F[x_1, \ldots, x_n]$ which is fixed only by the identity. What does Theorem 6-4 say in this case?

5. Let H be a subgroup of S_n and for the function f of the preceding problem, form

$$g = \prod_{\sigma \epsilon H} f_\sigma$$

Show that H is the subgroup fixing g.

Exercise 6-5

1. Let $\alpha_1,\alpha_2,\alpha_3,\alpha_4$ be the roots of

$$x^4 + px^2 + qx + r$$

and let H be the subgroup of S_4 fixing

$$\theta = (\alpha_1 - \alpha_2 + \alpha_3 - \alpha_4)^2$$

Show that H has order 8. Then find a cubic which has θ as one of its root.

2. Show that if $\alpha_1,\alpha_2,\alpha_3,\alpha_4$ are the roots of $x^4 + px^2 + qx + r$, then

$$(\alpha_1 + \alpha_2)(\alpha_1 + \alpha_3)(\alpha_2 + \alpha_4) = -q$$

3. Show that if the square roots of θ_1,θ_2, and θ_3 [where the θ_i are defined by Equation 6-19)] are chosen such that $\sqrt{-\theta_1}\sqrt{-\theta_2}\sqrt{-\theta_3} = -q$, then the values for $\alpha_i, i = 1, \ldots, 4$ given in Equation (6-21) are indeed roots of the equation $x^4 + px^2 + qx + r = 0$.

4. Using the relation of Equation (6-19) compare the discriminant of Equation (6-13) with the discriminant of Equation (6-18). Show the discriminant of the former is $16p^4r - 4p^3q^2 - 128p^2r^2 + 144pq^2r - 27q^4 + 256r^3$.

5. Having seen the method of solving the quartic equation, can you think of a plan of attack on the fifth degree equation?

Exercise 6-6

1. Verify that if the Lagrange resolvents θ and ψ of the cubic equation

$$f(x) = x^3 + px + q = 0$$

satisfy Equation (6-41), then the quantities $\alpha_1,\alpha_2,\alpha_3$ defined in Equation (6-27) are roots of $f(x)$.

7

Algebraic Geometry

7-1 Algebraic Varieties

A polynomial $f(x) = a_0 + a_1x + \ldots a_nx^n$ in one variable x with for instance real coefficients has only finitely many roots. A polynomial $f(x,y)$ of two variables can have infinitely many roots. In fact, its roots will usually lie along a curve in the real Euclidean plane. For example, the polynomial $f(x,y) = y - x^2$ vanishes at (a,a^2) for any real number a, and the totality of its roots is the parabola $y = x^2$.

The subject of algebraic geometry relates geometric properties of curves, such as the preceding parabola, to algebraic properties of the polynomials defining the curves. In this chapter, we shall point out some of the more basic aspects of this deep and difficult subject.

Throughout this chapter, F will denote a fixed, but arbitrary field. The set of all ordered pairs (a,b) of elements a,b of F will be denoted by F^2 — this is just the affine geometry coordinatized by the field F. More generally, the set of all ordered n-tuples $(a_1,a_2, \ldots, a_n)$ of $a_1,a_2, \ldots, a_n$ in F is called the *affine n-space* coordinatized by F and is symbolized by F^n. The ordered n-tuples $(a_1,a_2, \ldots, a_n)$ are called the *points* of F^n. A point $p = (a_1, \ldots, a_n)$ is called a *zero* of a polynomial $f(x_1, \ldots, x_n)$ of $F[x_1, \ldots, x_n]$ provided $f(p) = f(a_1, \ldots, a_n) = 0$. For example, a zero of $f(x_1,x_2,x_3) = x_1^2 + x_2^2 + x_3^2 - 1$ is $(0,0,1)$.

A set γ of points of F^n is called a *variety* (or *algebraic variety*) if there exist polynomials $f_1, \ldots, f_m$ in $F[x_1, \ldots, x_n]$ such that

$$f_1(p) = \ldots = f_m(p) = 0 \quad \text{for all } p \text{ in } \gamma \qquad \textbf{(7-1)}$$

and if

$$f_1(q) = \ldots = f_m(q) = 0 \qquad \textbf{(7-2)}$$

then q is in γ.

Thus, a variety is precisely the set of common zeros of a finite set of polynomials.* Often for convenience we shall write $\gamma = \gamma(f_1, \ldots, f_m)$ in order to exhibit the polynomials $f_1, \ldots, f_m$ which define the variety γ.

Example 7-1. The unit circle $\mathcal{C}$ with center at the origin is a variety in $F^2 : \mathcal{C} = \gamma(f)$, where $f(x,y) = x^2 + y^2 - 1$.

Example 7-2. The planes $x + y - 3z = 0$ and $3x + y + z = 0$ intersect in a line l. This line is a variety in $F^3 : l = \gamma(f,g)$, where $f(x,y) = x + y - 3z$ and $g(x,y) = 3x + y + z$.

Example 7-3. The entire affine n-space F^n is a variety. In fact, $F^n = \gamma(f)$, where $f = 0$.

Example 7-4. Each point $(a_1, \ldots, a_n)$ is a variety, for $(a_1, \ldots, a_n) = \gamma(f, \ldots, f_n)$, where $f_i(x_1, \ldots, x_n) = x_i - a_i$, $i = 1, \ldots, n$.

Example 7-5. The line segment $l : y = x$, $0 \leq x \leq 1$ (see Figure 7-1) is not a variety in F^2, where F is the real field. For while there are many polynomials which vanish along l, these polynomials also vanish at

---

*In fact, a variety may be defined as the common zeros of even an infinite set of polynomials. In either case, the two definitions are equivalent because of the Hilbert Basis Theorem.

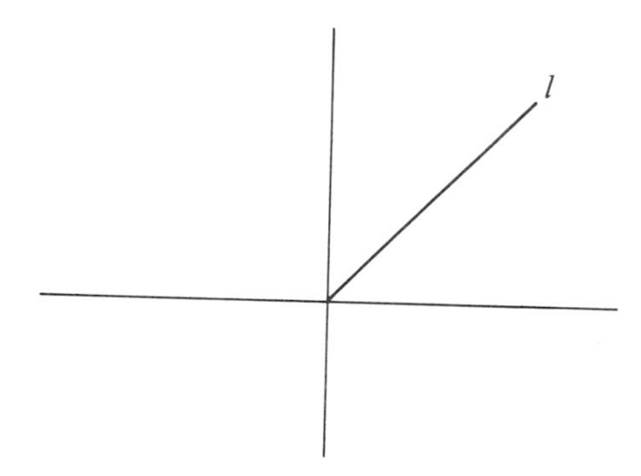

Figure 7-1

points outside l, contrary to the condition [Equation (7-2)] of the definition of variety. To see this, suppose $f(x,y)$ is a polynomial which vanishes along l. Then $f(x,x) = 0$ for $0 \leq x \leq 1$. Therefore, the polynomial $g(x) - f(x,x)$ has infinitely many roots. Consequently, $g(x) = 0$ for all x, and $f(x,y)$ vanishes along the entire line $y = x$.

It may happen that a variety can be decomposed into several smaller varieties. For example, in Figure 7-2 the curve γ consisting of the two

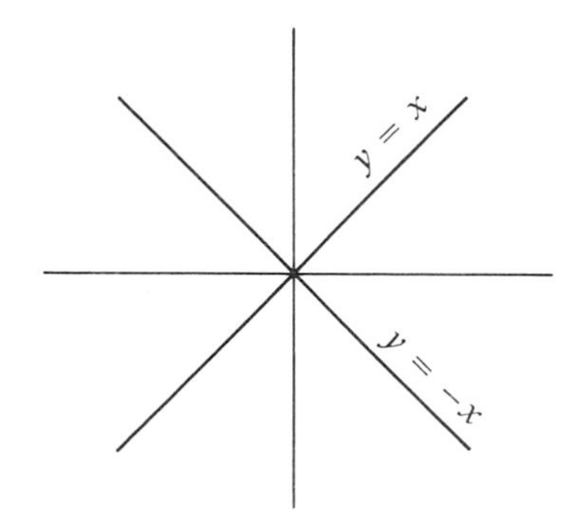

Figure 7-2

lines $y = x$ and $y = -x$ is a variety, for $\gamma = \gamma(x^2 - y^2)$. Also, γ is composed of the two smaller varieties $\gamma_1 = \gamma(x - y)$ and $\gamma_2 = \gamma(x + y)$.

Generally, a variety γ is called *reducible* if it is made up of smaller varieties $\gamma_1, \ldots, \gamma_t$. (We say that γ_i is smaller than γ in the sense that while γ_i is contained in γ, there is at least one point of γ which does not belong to γ_i.) Of course, we call γ *irreducible* if it is not reducible in the sense just mentioned.

7-2 The Coordinate Ring of a Variety

In this section we shall develop an algebraic criteria for deciding when a variety γ is irreducible.

Let γ be a variety in F^n and $f(x_1, \ldots, x_n), g(x_1, \ldots, x_n)$ be polynomials. We put

$$f(x_1, \ldots, x_n) \equiv g(x_1, \ldots, x_n) \pmod{\gamma}$$

providing f and g assume equal values on γ. That is, $f(x_1, \ldots, x_n) \equiv g(x_1, \ldots, x_n) \pmod{\gamma}$ if $f(a_1, \ldots, a_n) = g(a_1, \ldots, a_n)$ for each point $(a_1, \ldots, a_n)$ of γ.

Clearly, such a congruence defines an equivalence relation in the polynomial domain $F[x_1, \ldots, x_n]$. If $[f(x_1, \ldots, x_n)]$ denotes the equivalence class containing $f(x_1, \ldots, x_n)$, then addition and multiplication of these equivalence classes is defined by the following rules:

$$[f(x_1, \ldots, x_n)] + [g(x_1, \ldots, x_n)] = [f(x_1, \ldots, x_n) + g(x_1, \ldots, x_n)] \tag{7-3}$$

$$[f(x_1, \ldots, x_n)][g(x_1, \ldots, x_n)] = [f(x_1, \ldots, x_n)g(x_1, \ldots, x_n)] \tag{7-4}$$

It is not difficult to verify these rules are well defined and that the system of equivalence classes is a commutative ring with an identity element. This ring is called the *coordinate ring* of γ and is denoted by $F[x_1, \ldots, x_n]/(\gamma)$.

The geometric property of reducibility is reflected as an algebraic property of the coordinate ring as follows:

Theorem 7-1. The variety γ is irreducible if and only if its coordinate ring $F[x_1, \ldots, x_n]/(\gamma)$ is an integral domain.

For example, for the variety γ of Figure 7-2, the coordinate ring $F[x,y]/(\gamma)$ contains the nonzero elements $[x - y]$ and $[x + y]$. However, $[x + y][x - y] = [x^2 - y^2] = 0$ since $x^2 - y^2$ vanishes on γ. Thus, $F[x,y]/(\gamma)$ is not an integral domain.

Turning now to the proof of this theorem, let us suppose first that $F[x_1, \ldots, x_n]/(\gamma)$ is not an integral domain. Then, there exist $f(x_1, \ldots, x_n)$ and $g(x_1, \ldots, x_n)$ in $F[x_1, \ldots, x_n]$ such that the product $f(x_1, \ldots, x_n)g(x_1, \ldots, x_n)$ vanishes at each point of γ, but $f(p) \neq 0$ and $g(q) \neq 0$ for some p and q of γ.

Let γ_f denote the set of those points of γ at which f vanishes, and similarly let γ_g denote those zeros of g which lie on γ. Each of γ_f and γ_g is a variety. For if $\gamma = \gamma(h_1, \ldots, h_m)$ then $\gamma_f = \gamma_f(f,h_1, \ldots, h_m)$ and $\gamma_g = \gamma_g(g,h_1, \ldots, h_m)$.

Moreover, γ is composed of the two varieties γ_f and γ_g. For if p' is any point of γ, then $0 = f(p')g(p')$, hence, either $f(p') = 0$ or $g(p') = 0$. In the first case, p' lies on γ_f, while in the second case, p' lies on γ_g. We have shown that γ can be split into the varieties γ_f and γ_g, and each of these is actually smaller than γ itself since p is not on γ_f and q is not on γ_g.

Thus, if the coordinate ring of γ is not an integral domain then γ is reducible.

Conversely, assume γ is reducible. Then γ can be broken down into several smaller varieties. In fact, we shall assume that γ is composed of two varieties γ_1 and γ_2, each $\neq \gamma$. Of course, it may happen that γ is composed of three or more components. However, the argument in that case will not be essentially different from that we are about to give.

Since γ_1 and γ_2 are varieties, $\gamma_1 = \gamma_1(f_1, \ldots, f_m)$ and $\gamma_2 = \gamma_2(g_1, \ldots, g_k)$ for some $f_1, \ldots, f_m, g_1, \ldots, g_k$ in $F[x_1, \ldots, x_n]$. As γ_1 is smaller than γ, some point p of γ does not lie on γ_1; hence $f_i(p) \neq 0$ for some i, $1 \leq i \leq m$. Similarly, there is a point q of γ at which $g_j(q) \neq 0$ for some j, $1 \leq j \leq k$. Evaluating f_i and g_j at the points p and q, we see that $[f_i] \neq 0$ and $[g_j] \neq 0$. However, $[f_i g_j] = 0$. For if p' is any point of γ, then either p' lies on γ_1 or γ_2. In the first case, $f_i(p') = 0$, while in the latter, $g_j(p') = 0$. In either event, $f_i(p')g_j(p') = 0$. Therefore, the coordinate ring is not an integral domain, and this completes the proof of Theorem 7-1.

7-3 Finding the Points of a Variety

In principle, to find the points p belonging to a variety $\gamma = \gamma(f_1, \ldots, f_m)$, where $f_1, \ldots, f_m$ are in $F[x_1, \ldots, x_n]$, one solves the system

$$f_1(x_1, \ldots, x_n) = 0$$
$$\vdots \qquad \textbf{(7-5)}$$
$$f_m(x_1, \ldots, x_n) = 0$$

The points p of γ will be the solutions $p = (a_1, \ldots, a_n)$ of (7-5).

The system (7-5) may be solved by a process known as *Kronecker's method of elimination.** However, this method requires a great deal of computation and the explanation of its theory is much too lengthy to be given here. What we shall do instead is to make a few simple observations about (7-5) and show how the problem of solving (7-5) may be somewhat reduced.

To fix ideas, let us consider a variety $\gamma = \gamma(f,g)$ in the affine plane F^2. To find the points of γ we must solve

$$\begin{aligned} f(x,y) &= 0 \\ g(x,y) &= 0 \end{aligned} \qquad \textbf{(7-6)}$$

For the moment, think of x as fixed and regard f and g as polynomials in y alone. Using the division algorithm, we may write

$$f(x,y) = g(x,y)Q + R \qquad \textbf{(7-7)}$$

for some quotient Q and remainder R, where Q and R are polynomials in y and the degree of y in R is less than that of y in $g(x,y)$. Now the coefficients of Q and R depend on x. In fact, these coefficients may well be only rational functions of x. For example, if $g(x,y) = xy + 1$ and $f(x,y) = y^2 + x$, then by long division,

$$f(x,y) = g(x,y)\left(\frac{1}{x}y - \frac{1}{x^2}\right) + \left(\frac{1}{x^2} + x\right)$$

Clearing the denominators in Equation (7-7), we will have

$$A(x)f(x,y) = g(x,y)q(x,y) + r(x,y) \qquad \textbf{(7-8)}$$

for some $A(x)$ in $F[x]$ and $q(x,y)$, $r(x,y)$ in $F[x,y]$. [Doing this for the preceding example, $x^2f(x,y) = g(x,y)(xy - 1) + (1 + x^3)$]. We do not disturb the degree of y in the terms of Equation (7-7) when clearing the denominators from Q and R; hence, in Equation (7-8), $r(x,y)$ is still of

---

*See *Moderen Algebra*, vol II, by B.L. van der Waerden. Frederick Ungar Publishing Co., New York, 1949.

smaller degree in y than $g(x,y)$. Moreover, it follows from Equation (7-8) that any solution of Equation (7-6) is also a solution of

$$\begin{aligned} f(x,y) &= 0 \\ g(x,y) &= 0 \\ r(x,y) &= 0 \end{aligned} \tag{7-9}$$

It is nearly true that the solutions of the system

$$\begin{aligned} g(x,y) &= 0 \\ r(x,y) &= 0 \end{aligned} \tag{7-10}$$

give the solutions of Equation (7-6). For if (a,b) is a solution of (7-10), then substitution in Equation (7-8) gives $A(a)f(a,b) = 0$. Therefore, in reducing the system (7-6) to (7-10) we may say the following: If (a,b) is a solution of (7-10) and a is not among the roots of $A(x)$, then (a,b) is a solution of Equation (7-6). If a is a root of $A(x)$, then (a,b) may or may not be a solution of Equation (7-6). To decide whether it is or not, check the solutions of $h(y) = f(a,y) = 0$. If b is among the roots of $h(y)$, then (a,b) is a solution of Equation (7-6).

Summarizing, we have reduced the problem of solving (7-6) to the problem of solving (7-10) and finding the roots of polynomials in a single variable. The system (7-10) is actually simpler than (7-6) because the degree of y in $r(x,y)$ is smaller than that of $g(x,y)$.

The entire process is now repeated for (7-10). That is, using the division algorithm one replaces (7-10) with

$$\begin{aligned} r(x,y) &= 0 \\ r_1(x,y) &= 0 \end{aligned} \tag{7-11}$$

where $r_1(x,y)$ is a polynomial in $F[x,y]$ of y-degree less than that of $r(x,y)$. Specifically, we divide $r(x,y)$ into $g(x,y)$ to get a quotient and remainder and then clear denominators to get $r_1(x,y)$. [Of course, we may not divide by $r(x,y)$ if $r(x,y) = 0$; but in this case the system (7-10) would reduce to the single equation $g(x,y) = 0$ and its solution would be the variety $\gamma = \gamma(g)$].

Eventually, by such an Euclidean algorithm one may solve (7-6).

Example 7-5. Let us solve the system

$$\begin{aligned} y^3 + x^2y + 1 &= 0 \\ xy^2 + (x^3 + 1) &= 0 \end{aligned} \tag{7-12}$$

Dividing the second of these expressions into the first,

$$y^3 + x^2y + 1 = [xy^2 + (x^3 + 1)]\frac{1}{x}y + \left[\left(x^2 - \frac{x^3 + 1}{x}\right)y + 1\right]$$

Multiplying this equation by x,

$$x(y^3 + x^2y + 1) = [xy^2 + (x^3 + 1)]y + (-y + x)$$

Every solution of (7-12) will appear as a solution of the system

$$\begin{aligned} xy^2 + (x^3 + 1) &= 0 \\ -y + x &= 0 \end{aligned} \tag{7-13}$$

By inspection, the solutions of (7-13) are $x = y$ and $2x^3 + 1 = 0$. That is, if a,b, and c are the three cube roots of $-(1/2)$ then the solutions of (7-13) are (a,a),(b,b), and (c,c). It is easy to verify these are also solutions of (7-12).

Using the technique just outlined, we may determine precisely the irreducible varieties of the affine plane F^2.

The case where F has only finitely many elements is easily handled. For then F^2 will contain only finitely many points, as well as any variety in F^2. We have observed earlier that each point (a,b) of F^2 is a variety, for $(a,b) = \gamma(x - a, y - b)$. Consequently, if a variety γ of F^2 contains two or more points, then it can be decomposed into the smaller irreducible varieties consisting of its individual points. Therefore, if F is finite, then the only irreducible varieties of F^2 are those consisting of a single point.

Let us consider now the more interesting case, i.e., F is infinite. Assume γ is an irreducible variety of F^2. Then $\gamma = \gamma(f_1, \ldots, f_m)$ where $f_1, \ldots, f_m$ are in $F[x,y]$. If $f_1(x,y)$ is not an irreducible polynomial, then $f_1(x,y)$ factors, $f_1(x,y) = g_1(x,y)g_2(x,y)$, as a product of two polynomials of smaller degree. Passing to the coordinate ring $F[x,y]/(\gamma)$, $[g_1(xy)][g_2(x,y)] = [f_1(x,y)] = 0$. However, $F[x,y]/(\gamma)$ is an integral domain; hence, either $[g_1(x,y)] = 0$ or $[g_2(x,y)] = 0$. It is easy to see that in the first case $\gamma = \gamma(g_1, f_2, \ldots, f_m)$, while in the second, $\gamma = \gamma(g_2, f_2, \ldots, f_m)$.

Evidently by repeating such an analysis, we may eventually replace f_1 with one of its irreducible factors $\phi(x,y)$ without disturbing γ. That is,

$$\gamma = \gamma(\phi, f_2, \ldots, f_m)$$

for some irreducible polynomial $\phi(x,y)$ belonging to $F[x,y]$.

Now if ϕ divides each of the remaining polynomials $f_2, \ldots, f_m$, then $\gamma = \gamma(\phi)$. In this case γ has taken a relatively simple form. Let us suppose that ϕ does not divide f_2. The points of γ will be among the solutions of

$$\begin{aligned} \phi(x,y) &= 0 \\ f_2(x,y) &= 0 \end{aligned} \tag{7-14}$$

Consider x as fixed and ϕ and f_2 as polynomials in y alone. This means that ϕ and f_2 are regarded as polynomials from the domain D[y] where D = $F[x]$. The rational function field $K = F(x)$ is the quotient field of the integral domain D and Gauss's Lemma (and its consequences) tell us that ϕ is an irreducible polynomial of $K[y]$ and ϕ is not a factor of f_2 in $K[y]$. Thus, ϕ and f_2 are relatively prime in $K[y]$ and there exist $g(y)$ and $h(y)$ in $K[y]$ such that

$$1 = \phi(x,y)g(y) + f_2(x,y)h(y) \qquad \textbf{(7-15)}$$

The coefficients of $g(y)$ and $h(y)$ are rational functions of x, and if Equation (7-15) is multiplied by the least common multiple $J(x)$ of the denominators appearing in the coefficients of $g(y)$ and $h(y)$, then we will have

$$J(x) = \phi(x,y)G(x,y) + f_2(x,y)H(x,y) \qquad \textbf{(7-16)}$$

where $J(x)$ is in $F[x]$ and $G(x,y),H(x,y)$ belong to $F[x,y]$. Moreover, $J(x) \neq 0$. (The process we have used here is identical to that applied in passing from the system (7-6) to (7-8) previously.)

Now, if (a,b) is any point of γ, then $x = a$, $y = b$ is a solution of Equation (7-14). Substitution into Equation (7-16) shows that

$$J(a) = 0$$

However, $J(x)$ is a nonzero polynomial and has only finitely many roots. Consequently, if (a,b) belongs to γ, then a can assume only finitely many values. Thinking next of ϕ and f_2 as polynomials in x alone and repeating the preceding argument, we conclude there are only finitely many values for b as well.

Therefore, if ϕ does not divide each of $f_2, \ldots, f_m$, then $\gamma = \gamma(\phi, f_2, \ldots, f_m)$ contains only finitely many points. The irreducibility of γ forces γ to consist of only a single point.

The preceding analysis required only that $f_1 \neq 0$. If $f_1 = 0$ and some other f_i, $i \neq 1$, is nonzero, then we may repeat the analysis as applied to f_i rather than f_1. If it happens that $f_1 = \ldots = f_m = 0$, then $\gamma = \gamma(f_1, \ldots, f_m) = \gamma(0) = F^2$.

Summarizing our results, we have proved the following theorem:

Theorem 7-2. If γ is an irreducible variety of F^2, then γ is one of the following three types:

(1) $\gamma = F^2$

(2) $\gamma =$ a point

(3) $\gamma = \gamma(\phi)$, where $\phi(x,y)$ is an irreducibly polynomial of $F[x,y]$.

7-4 Tangents

If you were asked to describe in a naive way what it means for a line l to be tangent to a curve at a point P, you would most likely say that l touches the curve at P in such a way that if l is rotated slightly about P then it cuts the curve somewhere else, close to P (see Figure 7-3).

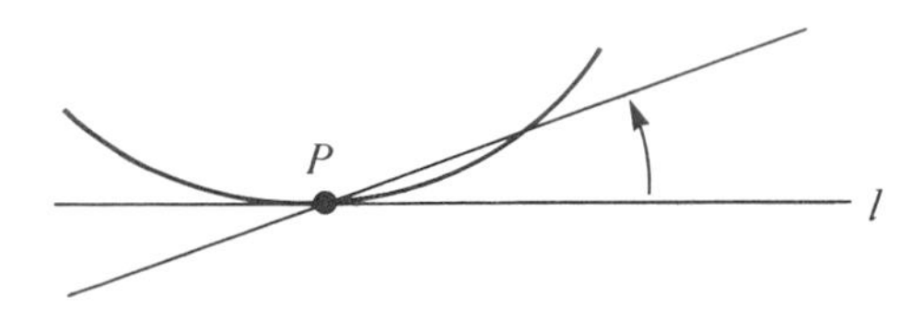

Figure 7-3

Of course, it is possible to describe l simply and precisely using calculus. However, let us pursue the naive description of tangent and see if it can be formulated in an analytic way. The reason for doing this is that we have already discussed curves in arbitrary affine planes F^2, where F need not be the real field, and for such fields F there may not be available a satisfactory theory of "limit" for determining tangents as in calculus. So our aim will be to express the notion of tangency in a *purely algebraic way*.

To simplify matters, let us choose $P = (0,0)$ to be the origin and suppose $\mathcal{C}$ is an algebraic curve passing through the origin. Thus, $\mathcal{C}$ will be the graph of the equation $\phi(x,y) = 0$, where $\phi(x,y)$ is a polynomial in x and y with coefficients from the field F. The polynomial $\phi(x,y)$ will have the form

$$\phi(x,y) = a+(bx + cy) + (dx^2 + exy + fy^2) + (gx^3 + hx^2y + \ldots) + \ldots$$

where we have grouped in parentheses terms of like degree and the coefficients $a,b,c,\ldots$ are in F. Since ϕ passes through (0,0), $a = 0$ and $\phi(x,y)= (bx + cy) + (dx^2 + exy + fy^2) + (gx^3 + hx^2y + \ldots) + \ldots$. Now, ϕ will contain only finitely many terms, and let us suppose that the highest (combined) degree in x and y of any of its terms is n. That is, $\phi(x,y)$ does contain some terms of the type $x^n, x^{n-1}y, n^{n-2}y^2, \ldots, y^n$, but no terms of higher degree.

To find the points at which the line $y = mx, m \in F$ intersects $\mathcal{C}$, we must solve the system

$$\phi(x,y) = 0$$
$$y = mx$$

or equivalently, find the roots of the polynomial $G_m(x)$ defined by $G_m(x) = f(x,mx)$. Writing out $G_m(x)$, we have

$$G_m(x) = (b + cm)x + (d + em + fm^2)x^2 + \ldots \qquad \textbf{(7-17)}$$

and the degree of $G_m(x)$ is $\leq n$.

For example, if $\phi(x,y) = x + 3y + xy + (x^2y - y^3)$ is of degree three, then $G_m(x) = (1 + 3m)x + mx^2 + (m - m^3)x^3$; hence, the degree of $G_m(x)$ is three if $m \neq 0,1$, or -1, while deg $G_m(x) = 1$ for $m = 0$, and deg $G_m(x) = 2$ for $m = \pm 1$.

From the form of Equation (7-17) of $G_m(x)$ we see that x is a factor of $G_m(x)$, which simply reflects the geometric fact that $y = mx$ cuts the curve $\mathcal{C}$ at the origin P. Now, it may happen that for some m the multiplicity of the root 0 of $G_m(x)$ is larger than one. If the multiplicity of the root 0 of $G_m(x)$ is r, then we say that the line $y = mx$ cuts the curve $\mathcal{C}$ with multiplicity r at P. In the example $\phi(x,y) = x + 3y + xy + (x^2y - y^3)$, the line $y = -1/3x$ cuts $\phi(x,y) = 0$ with multiplicity 2 at P, while $y = mx$, $m \neq -(1/3)$, cuts $\phi(x,y) = 0$ with multiplicity 1 at P.)

Now we define (algebraically) *tangent* as follows: Let r be the largest integer such that every line which cuts the curve $\phi(x,y) = 0$ at P cuts it with multiplicity r. Then a *tangent line* to $\phi(x,y) = 0$ at P is one which cuts the curve with multiplicity $> r$ at P.

Thus, for the curve $\phi(x,y) = x + 3y + xy + (x^2y - y^3) = 0$, $y = (-1/3)x$ is tangent to $\phi(x,y) = 0$ at (0,0). This result may be checked by calculus (for $F =$ the real field). Taking the differential $d\phi = 0$ and solving for dy/dx, we get

$$\frac{dy}{dx} = \frac{-1 - y - 2xy}{3 + x + x^2 - 3y^2}$$

At $x = y = 0$, $dy/dx = -1/3$ and the tangent line is $y = (-1/3)x$, as before.

For this example let us check that when the line $y = (-1/3)x$ is rotated slightly about the origin, it intersects the curve $\phi(x,y) = 0$ at another point near the origin. We had $G_m(x) = (1 + 3m)x + mx^2 + (m - m^3)x^3$ and $G_{-1/3}(x) = (-1/3)x^2 - (8/27)x^3$. If we rotate the tangent line slightly, then we obtain a line of slope n, which is approximately, but not equal to, $-1/3$. The polynomial $G_n(x)$ will still be of degree 3 because $n - n^3 \neq 0$. Therefore, $G_n(x)$ must have 3 roots. But now that 0 is no longer a multiple root of $G_n(x)$, it must have another root $\neq 0$. Geometrically this means that $G_n(x)$ has cut $\phi(x,y) = 0$ at another point different from (and near) the origin.

This analysis was a heuristic description of tangent line in the affine plane. It is deficient in two respects: (1) We restricted our study to what happens at the origin. (2) We dealt only with those lines of finite slope, whereas a curve may well have a vertical tangent line at a point. These two problems present no real difficulties, and we proceed now to give a final and precise formulation of tangent line.

Let $\phi(x,y) = 0$ be a curve in F^2 and $P = (a,b)$ be a point on this curve. Any line l passing through P will have parametric equations

$$x = a + \lambda t$$
$$y = b + \mu t$$

for some λ,μ in F, where t is the parameter of the line. To find the points at which l intersects $\phi(x,y) = 0$, we solve

$$\phi(x,y) = 0$$
$$x = a + \lambda t$$
$$y = b + \mu t$$

or what amounts to the same thing,

$$0 = \phi(a + \lambda t, b + \mu t) \qquad \textbf{(7-18)}$$

The right-hand side of Equation (7-18) is a polynomial $G_\ell(t)$, and $t = 0$ is a root of $G_\ell(t)$. As before, the line l is said to cut the curve $\phi(x,y) = 0$ with multiplicity r at P if r is the multiplicity of the root 0 of the polynomial $G_\ell(t)$. A *tangent* line is one which cuts $\phi(x,y) = 0$ with multiplicity $> r$, where r is the largest integer with the property that all lines intersecting $\phi(x,y) = 0$ at P cut the curve there with multiplicity r.

If every line cuts the curve $\phi(x,y) = 0$ at a point P with multiplicity ≥ 2, then the curve might not have a unique tangent at P. A simple example is the curve $\phi(x,y) = x^2 - y^2 = 0$ (see Figure 7-4). At (0,0) there are two distinct tangents. For a line $y = mx$ the corresponding polynomial $G_m(x) = \phi(x,mx) = (1 - m^2)x^2$ has a double root at $x = 0$ and the lines which cut $x^2 - y^2 = 0$ with maximal multiplicity at (0,0) are $y = x$ and $y = -x$ (corresponding to $m = 1$ and $m = -1$).

Generally, if $0 = \phi(x,y) = (ax + by) + (cx^2 + dxy + ey^2) + \ldots$ is a curve passing through the origin (0,0) and if every line through (0,0) cuts the curve with multiplicity ≥ 2 at the origin, then for each value of m, $\phi(x,mx) = (a + bm)x + (c + dm + em^e)x^2 + \ldots$ is divisible by x^2.

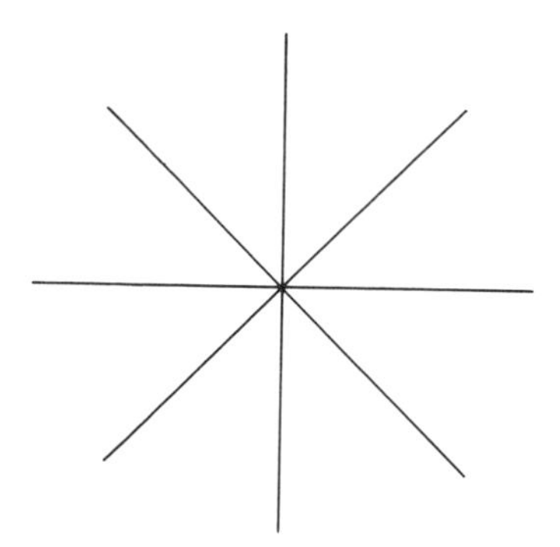

Figure 7-4

Thus, $a = 0$, $b = 0$, and $\phi(x,y) = (cx^2 + dxy + ey^2) + \ldots$. Assuming not all of c,d, and e are zero, there will be at most a pair of lines which cut $\phi(x,y) = 0$ three times at (0,0). For the polynomial $\phi(x,mx) = (c + dm + em^2)x^2 + \ldots$ is divisible by x^3 only if m is a root of the quadratic $c + dt + et^2$. For the real field this quadratic will have 1, 2, or no roots, depending on whether $d^2 - 4ec$ is zero, positive, or negative respectively, and corresponding to these cases there will be 1, 2, or no tangents to $\phi(x,y) = 0$ at (0,0). Pictorially, these cases are shown by Figure 7-5. Going from left to right in Figure 7-5, the origin is called a *node*, *cusp*, and *isolated point* respectively.

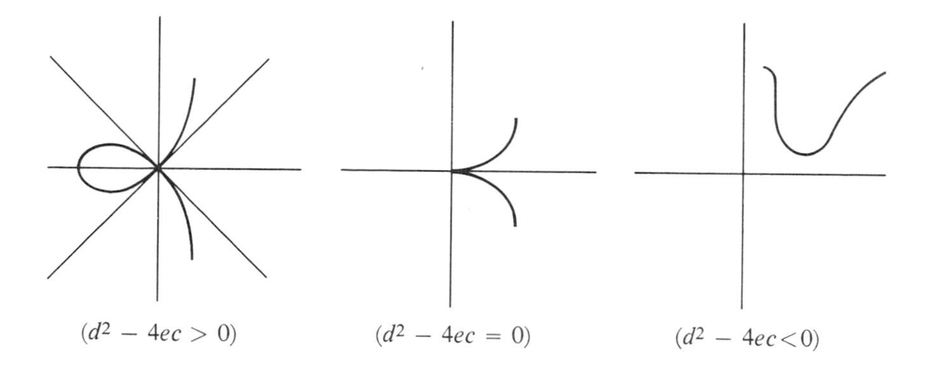

Figure 7-5

Naturally, the phenomena may occur at points other than the origin. Algebraic geometers refer to the preceding three types of points as *multiple* points, and they call the others, *simple* points. Algebraically, P is a simple point of $\phi(x,y) = 0$ if some line at least cuts $\phi(x,y) = 0$ only once at P.

7-5 Projective Geometry

With the techniques developed thus far, we may say a great deal about curves $\phi(x,y) = 0$ using just algebraic methods. For example, irreducibility may be checked using the coordinate ring, while tangents and multiple points are determined by calculating multiplicities of roots of polynomials. As well, x-intercepts and y-intercepts can be found by solving $\phi(x,0) = 0$ and $\phi(0,y) = 0$. We close this chapter with an investigation of the asymptotes to a curve $\phi(x,y) = 0$.

The procedure we shall follow is instructive and typical of mathematics: To fully comprehend a concrete problem, such as finding asymptotes to a curve, we try to picture the given situation in an "ideal" and imaginative setting. *By necessity, this setting is an abstract one.* In this general framework the problem is solved and the results are brought back "to reality."

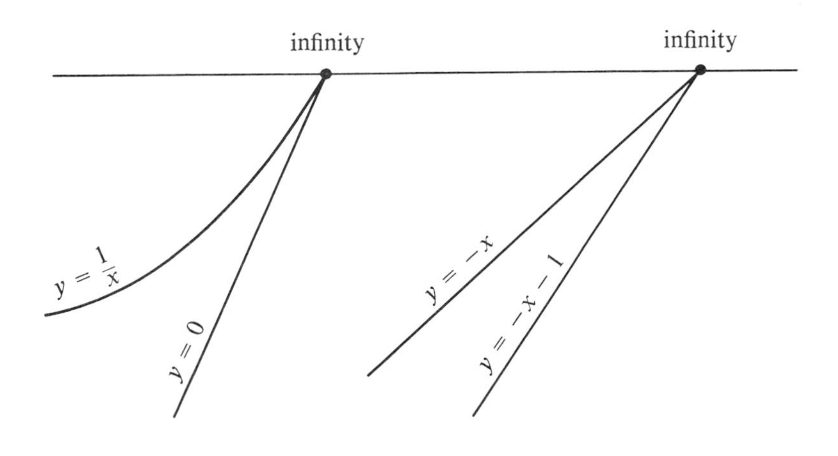

Figure 7-6

Specifically, an asymptote to a curve $\phi(x,y) = 0$ is a line which is tangent to the curve at *infinity.* We already know how to calculate tangents at ordinary points, so if we could coordinatize the plane so that "infinity" also has coordinates, then we would know how to find the asymptotes.

Consider Figure 7-6. If we were standing in the plane it would appear to us that the curve $y = 1/x$ and the line $y = 0$ meet on the horizon, i.e., at infinity. Moreover, any pair of parallel lines, such as $y = -x$ and $y = -x - 1$ meet at infinity but at a different point than where $y = 1/x$ and $y = 0$ meet. It is tempting to call the horizon the *line at infinity*, noting that all lines which are parallel will intersect at some point on this line at infinity.

If we consider the system $\mathcal{P}$ of all ordinary lines and points of the plane together with l_∞, the line at infinity, and all its points, then the geometric system $\mathcal{P}$ will have the following properties:

Property 7-1. For any pair of distinct points of $\mathcal{P}$ there is a unique line containing them.

Property 7-2. Each pair of distinct lines of $\mathcal{P}$ intersect in exactly one point.

A system of points and lines satisfying (5.2) and 5.3) is called a *projective geometry*, or a *projective plane.* A projective geometry differs from an affine geometry in that there are no parallel lines in a projective geometry.

A projective geometry $\mathcal{P}$ may be constructed from an affine geometry Π by adding to Π the line at infinity. For example, consider the affine geometry Π of Section 2-5 consisting of the points A,B,C,D and the lines

$$l_1 = \{A,B\} \qquad l_2 = \{C,D\} \qquad l_3 = \{A,D\}$$
$$l_4 = \{B,C\} \qquad l_5 = \{B,D\} \qquad l_6 = \{A,C\}$$

In $\mathcal{P}$ the parallel lines l_1 and l_2 must intersect at a point at infinity—call this new point E. Then l_3 and l_4, which are parallel, must intersect at a new point F (and $F \neq E$; otherwise, Property 7-2 would be violated). Finally, l_5 and l_6 will intersect at a point G. Setting $l_\infty = \{E,F,G\}$ we have the projective geometry $\mathcal{P}$ consisting of the points A,B,C,D,E,F,G and the lines $l_1' = \{A,B,E\}$, $l_2' = \{C,D,E\}$, $l_3' = \{A,D,F\}$, $l_4' = \{B,C,F\}$, $l_5' = \{B,D,G\}$, $l_6' = \{A,C,G\}$, and $l_\infty = \{E,F,G\}$.

7-6 Homogeneous Coordinates

We have explained how an affine geometry may be completed to a projective geometry in such a way that lines which were parallel in the affine geometry meet at infinity. Now we consider the problem of introducing "Cartesian" coordinates to the projective geometry. A naive way of accomplishing this would be to use ordinary coordinates to describe the points of the affine geometry and the symbol ∞ to denote the points at infinity. However, this is unsatisfactory on two counts: (1) There is more than one point at infinity. (2) The arithmetic involving ∞ is too artificial to be of any use.

To coordinatize a projective geometry in a suitable way, what is done is to use *three* coordinates rather than two to locate the points of the plane. It will be this extra coordinate which allows an analytic description of the points at infinity.

If (x,y) is a point of the affine plane, then *homogeneous* coordinates of this point are defined to be the triple (X,Y,Z), where $x = X/Z$, $y = Y/Z$. For example, homogeneous coordinates of $(2,3)$ are $(4,6,2)$. Given homogeneous coordinates (X,Y,Z) of a point we find its ordinary coordinates (x,y) from the equations $x = X/Z$, $y = Y/Z$, if $Z = 0$. It should be noted that a point (x,y) may be described in many different ways by homogeneous coordinates. For example, $(4,6,2),(-2,-3,-1)$ and $(7,21/2,7/2)$ all represent the point $(2,3)$. Generally, (X,Y,Z) and (X',Y',Z') are said to represent the same point if $X/Z = X'/Z'$ and $Y/Z = Y'/Z'$, and this occurs when $X = kX',Y = kY',Z = kZ'$ for some $k \neq 0$.

Now to obtain the ordinary coordinates of a point in the affine plane from its homogeneous coordinates (X,Y,Z), we divide by Z; and this is possible only if $Z \neq 0$. That is, the homogeneous point (X,Y,Z) corresponds to an affine (ordinary) point only if $Z \neq 0$. It is natural to say that the points of the form $(X,Y,0)$ correspond to the points at infinity. As shown previously, $(X,Y,0)$ and $(X',Y',0)$ represent the same point at infinity if and only if $X = kX'$ and $Y = kY'$ for some $k \neq 0$. The line at infinity contains those points of the type $(X,Y,0)$. In homogeneous coordinates no meaning is attached to the triple $(0,0,0)$ as this would correspond to the indeterminate forms $x = 0/0$, $y = 0/0$.

Summarizing, the projective plane $\mathcal{P}$ coordinatized by a field F consists of ordered triples (X,Y,Z), where X,Y, and Z are not all zero and belong to F. Triples (X,Y,Z) and (X',Y',Z') are called *equal* if $X = kX',Y = kY'$, and $Z = kZ'$ for some $k \neq 0$ in F. The point (X,Y,Z), $Z \neq 0$, is really the ordinary point $(X/Z,Y/Z)$ and l_∞, the line at infinity, is given by

$$l_\infty = \{(X,Y,0)\}$$

Example 7-6. Find the points at which the hyperbola $y = 1/x$ intersects the line at infinity.

Solution: What is meant is to find the homogeneous coordinates of the points at infinity which lie on $xy = 1$. Using the transformations $x = X/Z$ and $y = Y/Z$, the equation $xy = 1$ in homogeneous form becomes $(X/Z)(Y/Z) = 1$, or $XY = Z^2$. Now the hyperbola $XY = Z^2$ cuts l_∞ when $Z = 0$; hence, when $X = 0$ (and $Y \neq 0$) or when $Y = 0$ (and $X \neq 0$). Thus, the points at infinity on the hyperbola are $(0,Y,0)$ and $(X,0,0)$, where $X \neq 0$ and $Y \neq 0$. But, $(0,Y,0)$ and $(0,1,0)$ represent the same point, as do $(X,0,0,)$ and $(1,0,0)$. Therefore, we may say that $xy = 1$ intersects l_∞ at $(1,0,0)$ and $(0,1,0)$.

Example 7-7. The circle $x^2 + y^2 = 1$ in the real plane does not intersect l_∞ (of course). This may be shown using homogeneous coordinates. Making the transformations $x = X/Y$, $y = Y/Z$, and clearing denominators, the circle becomes $X^2 + Y^2 = Z^2$, in homogeneous coordinates. When $Z = 0$, $X = Y = 0$. But, $(0,0,0)$ does not represent any point in projective geometry. Hence, the circle has no point at infinity.

7-7 Asymptotes

In calculus the asymptotes (if any) of the curve $\phi(x,y) = 0$ are found by expressing y as a function of x (or x as a function of y) and then taking limits as $x \longrightarrow \infty$. Such a procedure works only in a limited number of cases, because it requires that from the equation $\phi(x,y) = 0$ one of the variables may be expressed as a function of the other.

Our program for calculating the asymptotes of $\phi(x,y) = 0$ is entirely different, and does not require limits. It is based on the principle that an asymptote of $\phi(x,y) = 0$ is tangent to the curve at infinity. The process itself consists of three steps:

Step 1. Find the points at which $\phi(x,y) = 0$ cuts l_∞, the line at infinity.
Step 2. Find those lines which intersect $\phi(x,y) = 0$ at infinity.
Step 3. Of these lines, find those which cut $\phi(x,y) = 0$ at infinity with maximum multiplicity — these will be the asymptotes.

Remark 7-1. Step 1 is accomplished by expressing the curve $\phi(x,y) = 0$ in homogeneous form by replacing x with X/Z and y with Y/Z and then eliminating Z from the denominators by multiplying by a suitable power of Z. Then the points at infinity are found by setting $Z = 0$.

In Step 2, once a given line intersects $\phi(x,y) = 0$ at infinity, any line parallel to this line will also intersect $\phi(x,y) = 0$ at the same point of infinity. In this way one obtains many candidates for the tangents. The actual tangents are sifted from these lines by calculating the multiplicities of the intersections.

It is time now to consider a concrete example. Let us find the asymptotes of the hyperbola

$$\frac{x^2}{3} - \frac{y^2}{4} = 1$$

done over the real field. In homogeneous coordinates the curve is

$$\frac{X^2}{3} - \frac{Y^2}{4} = Z^2$$

and it intersects l_∞ when $Z = 0$ and $(2X - \sqrt{3}Y)(2X + \sqrt{3}Y) = 0$. Thus, homogeneous coordinates of the two points at infinity of this curve are

$$\left(1,\frac{2}{\sqrt{3}},0\right) \quad \text{and} \quad \left(1,\frac{-2}{\sqrt{3}},0\right)$$

where we have taken (without loss of generality) the value $X = 1$. Evidently these are distinct points since the coordinates of the above pair are not multiples of one another.

Looking at one branch of the hyperbola, we have Figure 7-7.

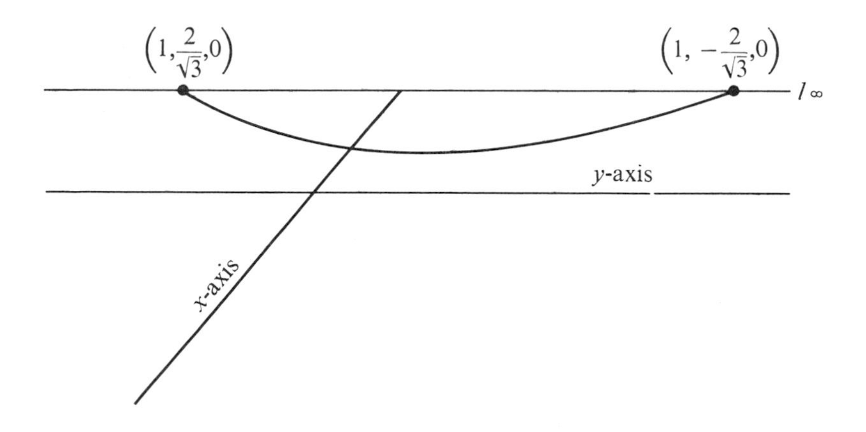

Figure 7-7

We find now those lines l which intersect l_∞ at $(1,2/\sqrt{3},0)$. Such a line l will have the equation $ax + by - c = 0$, where a and b are not both zero. The homogeneous equation for l is $aX + bY - cZ = 0$ and l intersects l_∞ at $(b,-a,0)$. For $(b,-a,0)$ and $(1,2/\sqrt{3},0)$ to represent the same point,

$$b = k \cdot 1$$

$$-a = k \cdot \frac{2}{\sqrt{3}}$$

for some $k \neq 0$. The preceding relations may be satisfied by choosing $b = 1$, $k = 1$, and $a = -2/\sqrt{3}$. Thus, the parallel lines

$$l: \left(\frac{-2}{\sqrt{3}}\right)x + y - c = 0 \qquad \textbf{(7-19)}$$

all intersect the hyperbola at $(1,2/\sqrt{3},0)$. Since lines intersecting l_∞ at $(1,2/\sqrt{3},0)$ are necessarily parallel, we have found *all* lines intersecting l_∞ at $(1,2/\sqrt{3},0)$. See Figure 7-8.

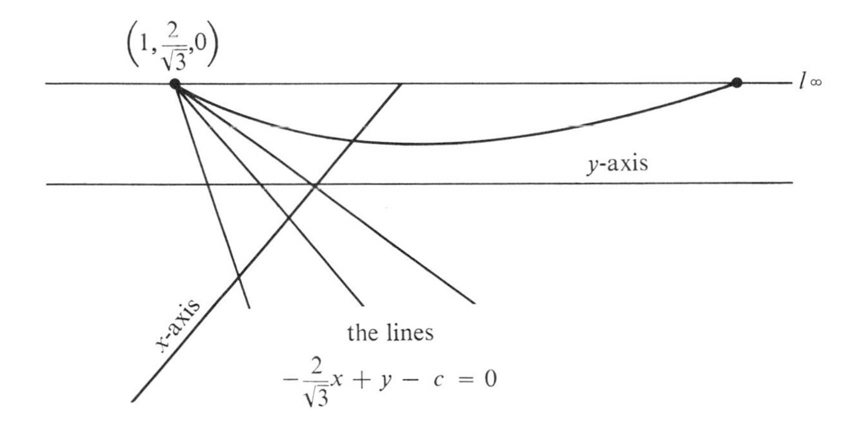

Figure 7-8

Now we have to decide which of the lines of Equation (7-19) cut the hyperbola with maximal multiplicity at $(1,2/\sqrt{3},0)$. In homogeneous coordinates a line of Equation (7-19) and the hyperbola have equations

$$\frac{X^2}{3} - \frac{Y^2}{4} = Z^2$$

$$-\left(\frac{2}{\sqrt{3}}\right)X + Y - cZ = 0 \qquad \textbf{(7-20)}$$

Eliminating Y, $(X^2/3) - (1/4)[(2/\sqrt{3})X + cZ]^2 = Z^2$, or

$$Z\left[\frac{cX}{\sqrt{3}} + \left(1 + \frac{c^2}{4}\right)Z\right] = 0 \qquad \textbf{(7-21)}$$

When $X = 1$, $Z[c/\sqrt{3} + (1 + c^2/4)Z] = 0$. That is, 0 is a root of $G(Z) = Z[c/\sqrt{3} + (1 + c^2/4)Z]$, reflecting the fact that the lines and hyperbola intersect when $X = 1$, $Y = 2/\sqrt{3}$, $Z = 0$. The line which is asymptotic to the hyperbola corresponds to that value of c for which 0 is a root of maximum multiplicity of $G(Z)$. Clearly this occurs when $c = 0$ (whence, $G(Z) = Z^2$). Thus,

$$\frac{-2}{\sqrt{3}}x + y = 0 \qquad \textbf{(7-22)}$$

is asymptotic to $(x^2/3) - (y^2/4) = 1$.

In a similar way one finds the line asymptotic to the hyperbola at the other point of infinity.

Exercises

Exercise 7-1

1. Show that the curve parametrized by the equations

$$x = t$$
$$y = t^2 + 1$$
$$z = t^3 - t$$

is an algebraic variety of F^3, where F is any field.

2. Show that the helix

$$x = \cos t$$
$$y = \sin t$$
$$z = t$$

is not an algebraic variety of F^3, where F is the real field.

3. For the real field F show that if $\gamma = \gamma(f,g)$ is a variety of F^2 then there exists a polynomial $h(x_1,x_2) \in F[x_1,x_2]$ such that $\gamma = \gamma(h)$.

4. Show that the result of Problem 3 is false for the complex field F.

5. Show that if F is a field with only finitely many elements, then any set of points of F^n is an algebraic variety.

6. Let $f,g \in F[x_1, \ldots, x_n]$, where F is any field. Prove that $\gamma(f,g) = \gamma(f) \cap \gamma(g)$ and $\gamma(fg) = \gamma(f) \cup \gamma(g)$.

Exercise 7-2

1. For a field F let γ be an irreducible variety of F^n, and let K be the quotient field of the integral domain $F[x_1, \ldots, x_n]/(\gamma)$.
   (a) Show that K is an extension of F.
   (b) Show that K consists of all quotients f/g, where $f,g \in F[x_1, \ldots, x_n]$ and $g(x_1, \ldots, x_n) \not\equiv 0 \pmod{\gamma}$.

2. Let $\gamma = \gamma(y - x^2)$.
   (a) Show that γ is an irreducible variety of F^2, where F is any field.
   (b) Show that the quotient field of the coordinate ring of γ is isomorphic to $F(x)$.

3. Let γ be a point of F^n, where F is any field. Show that the coordinate ring of γ is just F.

Exercise 7-3

1. Show that if $f(x_1,x_2) \in F[x_1,x_2]$ is irreducible and $\gamma(f)$ is infinite, then $\gamma(f)$ is irreducible. (This is a partial converse of Theorem 7-2.)

2. Give an example of a polynomial $f(x_1,x_2) \in F[x_1,x_2]$, where F is the real field, such that $f(x_1,x_2)$ is irreducible but $\gamma(f)$ is not irreducible.

3. Describe the points which belong to the variety $\gamma(y^2 + x^3, y^3 + xy^2 + x^3)$.
4. Describe the points which belong to the variety $\gamma(y^5 - x^2y, y^4 + (x - x^2), y^2 - x^3)$.

5. Decompose each of the following varieties into their irreducible parts.
   (a) $\gamma(y^2 - xy - x^2y + x^3)$
   (b) $\gamma(x^2y^2 + xy + x^3y + x^2y + x^2 + x)$
   (c) $\gamma(x^4y^2 - x^5y + xy^2 - x^3y - y + x^4 - x^2y + x)$

Exercise 7-4

1. Show that if $f(x,y) \in F[x,y]$, where F is any field, then $\gamma(f)$ has only finitely many multiple points.

2. Let $f(x,y) = (x + y)(2x + y)(3x + y) + x^4y^7 + x^6y^3$. Find the tangents to the curve $f(x,y) = 0$ at the origin.

3. Describe the multiple point of $y^3 - x^2 = 0$ at the origin.

4. Give an example of a curve which has a node at (1,1).

5. Give an example of a curve which has an isolated point at (1,1).

Exercise 7-5

1. Let Π be the affine geometry coordinatized by the field $I/(3)$. Extend Π to a projective geometry.

2. An *arc* of a projective geometry is a set of 4 or more points, no 3 of which are collinear. Find an arc in the projective geometry of problem 1. Give an example of a projective geometry which has no arcs. Show that if $\mathcal{P}$ is a projective geometry containing an arc and a line with $n + 1$ points, where $n \geq 1$, then all lines of $\mathcal{P}$ will contain exactly $n + 1$ points. Show also that the geometry $\mathcal{P}$ must contain exactly $n^2 + n + 1$ points and $n^2 + n + 1$ lines.

Exercise 7-6

1. Find the points at infinity on each of the following curves of F^2, where F is the real field.
   (a) $y = x^2$
   (b) $y^2 = x^3$
   (c) $y^3 + x^2y + xy^2 + 1 = 0$

2. Write each of the following curves of F^2, where F is the real field, in homogeneous coordinates.
   (a) $y = mx + b$
   (b) $2x^2 + 3y^2 = 1$
   (c) $x^2 - y^2 = 2$

3. Consider the ellipse $(1/2)x^2 + y^2 = 1$ of F^2. Find the points at infinity which lie on this ellipse for each of the following cases.
   (a) $F =$ the complex field
   (b) $F =$ the real field
   (c) $F =$ the rational field
   (d) $F = I/(5)$

Exercise 7-7

1. Find the asymptotes of the hyberbola

$$\frac{x^2}{a^2} - \frac{y^2}{b^2} = 1$$

   where $a,b \in F$, the real field.

2. What does it mean for the curve $f(x,y) = 0$ to be tangent to the line at infinity? Consider, for example, the curve $y = x^2$.

3. Find the asymptotes, if any, of the following.
   (a) $x^2y + x + y^3 = 1$
   (b) $y^2 + x^3y^2 + x^2y^3 = 1$

4. Consider the circle $x^2 + y^2 = 1$ of F^2, where F is the complex field. Find the asymptotes, if any, of this circle.

Appendix

Sets

If S is the set of all objects which have in common a certain property P, then we shall indicate this by the notation

$$S = \{x \mid x \text{ has property } P\} \tag{1}$$

The vertical line in the symbol $\{\ \mid \ldots\}$ should be read as "such that". For example,

$$S = \{x \mid 0 \leq x \leq 1\}$$

denotes the set of numbers x such that $0 \leq x \leq 1$, while

$$\{x \mid x \text{ is an odd integer}\}$$

is the set of integers $\pm 1, \pm 3, \pm 5, \ldots$.

If an object x belongs to a set S, then this is indicated by the notation $x \in S$. On the other hand, if x does not belong to S, then this is described by the notation $x \notin S$.

If the elements of a set S are distributed among various subsets $A_1, A_2, \ldots, A_n$, then we write

$$S = A_1 \cup A_2 \cup \ldots \cup A_n \tag{2}$$

to show this. For example, if S is the set of all integers and

$$A_1 = \{x \in S \mid x \leq 2\}$$
$$A_2 = \{x \in S \mid 1 \leq x \leq 3\}$$
$$A_3 = \{x \in S \mid x \geq 1\}$$

then, $S = A_1 \cup A_2 \cup A_3$. Similarly, if S is the set of all even integers, A is the set of all even integers, and B is the set of all odd integers, then $S = A \cup B$. In this example, the subsets A and B have no integers, i.e., they do not overlap. However, in the first example, there are integers which belong to each of the subsets A_1, A_2, and A_3.

If A and B are sets of objects, then $A \cap B$ denotes the set of objects which simultaneously belongs to A and B. For example, if $A = \{x \mid 2 \leq x \leq 4\}$ and $B = \{x \mid -(1/2) \leq x \leq 3\}$, then $A \cap B = \{x \mid 2 \leq x \leq 3\}$. More generally, if $A, B, C, \ldots$ are sets, then $A \cap B \cap C \cap \ldots \cap$ is the set of objects which belong to each of $A, B, C, \ldots$.

Logic

If P and Q are statements then the notation

$$P \Rightarrow Q \tag{3}$$

means that the statement Q is a logical consequence of the statement P. For example, in calculus, functions which can be differentiated are continuous. In terms of Equation (3), this could be written as

$$f \text{ differentiable} \Rightarrow f \text{ continuous}$$

For statements P and Q the notation

$$P \Leftrightarrow Q$$

means that P and Q are logically equivalent statements, i.e., $P \Rightarrow Q$ and $Q \Rightarrow P$.

Sums and Products

If $a_1, a_2, \ldots, a_n$ are numbers, then their sum $a_1 + a_2 + \ldots + a_n$ is commonly denoted by the symbol Σ. Thus,

$$a_1 + a_2 + \ldots + a_n = \sum_{k=1}^{k=n} a_k$$

Occassionally, one writes

$$a_1 + a_2 + \ldots + a_n = \sum_{1 \leq k \leq n} a_k$$

for the sum.

The product $a_1 a_2 \ldots a_n$ of $a_1, a_2, \ldots, a_n$ is written as

$$a_1 a_2 \ldots a_n = \prod_{k=1}^{k=n} a_k$$

or

$$a_1 a_2 \ldots a_n = \prod_{1 \leq k \leq n} a_k$$

Index